California Cures!

How the California Stem Cell Program is Fighting Your Incurable Disease!

California Cures!
How the California Stem Cell Program is Fighting Your Incurable Disease!

Don C. Reed
Americans for Cures Foundation, USA

World Scientific

NEW JERSEY · LONDON · SINGAPORE · BEIJING · SHANGHAI · HONG KONG · TAIPEI · CHENNAI · TOKYO

Published by

World Scientific Publishing Co. Pte. Ltd.

5 Toh Tuck Link, Singapore 596224

USA office: 27 Warren Street, Suite 401-402, Hackensack, NJ 07601

UK office: 57 Shelton Street, Covent Garden, London WC2H 9HE

Library of Congress Cataloging-in-Publication Data

Names: Reed, Don C., author.

Title: California cures! : how the California stem cell program is fighting your incurable disease! / by Don C. Reed, Americans for Cures Foundation, USA.

Description: New Jersey : World Scientific, [2018] | Includes bibliographical references and index.

Identifiers: LCCN 2017057973| ISBN 9789813231368 (hardcover : alk. paper) | ISBN 981323136X (hardcover : alk. paper)

Subjects: LCSH: Stem cells--Therapeutic use. | Stem cells--Research--California.

Classification: LCC QH588.S83 R43 2018 | DDC 616.02/774--dc23

LC record available at https://lccn.loc.gov/2017057973

British Library Cataloguing-in-Publication Data

A catalogue record for this book is available from the British Library.

First published 2018 (Hardcover)

Reprinted 2019 (in paperback edition)

ISBN 978-981-3270-38-1 (pbk)

Copyright © 2018 by World Scientific Publishing Co. Pte. Ltd.

For photocopying of material in this volume, please pay a copying fee through the Copyright Clearance Center, Inc., 222 Rosewood Drive, Danvers, MA 01923, USA. In this case permission to photocopy is not required from the publisher.

For any available supplementary material, please visit

https://www.worldscientific.com/worldscibooks/10.1142/10747#t=suppl

To all who suffer: may we work together, and defeat the cause of your pain

CONTENTS

Introduction: Evangelina and the Golden State — 1

1 The Absolute Minimum You Need to Know First — 5
2 To Breathe, or Not to Breathe — 9
3 The Strongest Man in the World — 13
4 When the Dolphin Broke My Ear — 17
5 The Boy with Butterfly Skin — 23
6 The Great Baldness "Comb-Over" Replacement? — 29
7 "He Sees! He Sees!" — 33
8 Cop at the Window — 39
9 "Go West, Young (Wo)Man" — To a Biomed Career? — 43
10 And How Will You be Paying for that New Heart? — 49
11 The Answer to Cancer? — 53
12 A Political Obstacle to Heart Disease Cure? — 59
13 Your Friend, the Liver! — 65
14 "Bring 'em Back Alive" — 69
15 The Color of Fat — 75
16 Revenge for My Sister — 81
17 A Story with No Happy Ending? — 87
18 Aging and Stem Cells — 93
19 The "Impending Alzheimer's Healthcare Disaster" — 97
20 President Trump's Great Stem Cell Opportunity — 103

21	Leiningen's Ants and Parkinson's Disease	107
22	On the Morality of Fetal Cell Research	111
23	Democracy and Gloria's Knees	115
24	Three Children, and the Eternal Flame	121
25	Autism, Mini-Brains, and the Zika Virus	127
26	Why "The Big Bang Theory" Matters to Me	131
27	Musashi and the Two-Sword Solution	135
28	"The Magnificent Seven"	139
29	The Connecticut Commitment	145
30	In Memory of Beau	149
31	To Relocate Alligators, or Turn a Country on to Biomed?	155
32	Whale Sharks and Outer Space	161
33	Mr. Science Goes to Washington?	169
34	When Oklahoma Is Not Ok	173
35	James Bond and Melanoma	179
36	Neurological Diseases vs. California	183
37	Driving to the Storm	187
38	Door into Tomorrow	193
39	Stem Cell Battles — On Times Square?	203
40	Annette, Richard Pryor, and Multiple Sclerosis	209
41	Mike Pence, and Reproductive Servitude	215
42	Motorcycle Wrecks and Complex Fractures	219
43	Even Dracula Gets Arthritis	225
44	Tugboat for Cure	229
45	Wheelchair Warriors, Take Back Your Rights!	235
46	Sickle Cell Disease vs. Stem Cell Agency	239
47	Dwight Clark, "The Catch," and A.L.S.	245
48	A Friend is Lost	249
49	Dying in Doonesbury, Fighting Back at UCD	253
50	The Man with the Autographed Baseball	257

51   The Gorilla Gynecologist Returns   259
52   Wrestling the Invisible Enemy   263
53   Two Warriors Named Joan   267
54   An End to Heroism?   273
55   Message from the Middle Kingdom   275
56   Scientists and the Undocumented   281
57   The Girl, the Bandit, and Women in Science   285
58   The Greatest Proposal   293
59   Forty-Two California Clinical Trials   303
60   Gathering of Champions   307
61   Goodbye, Hello!   317
62   The Answer   323
63   A Nobel Prize for Bob Klein?   325
Afterword: For More Information   331
Personal Message   335
Index   337

INTRODUCTION: EVANGELINA AND THE GOLDEN STATE

Once there was a baby girl, just one-year old, pretty a baby as ever was born. Her mom and dad loved her, of course — how could they not?

But there was a problem.[1]

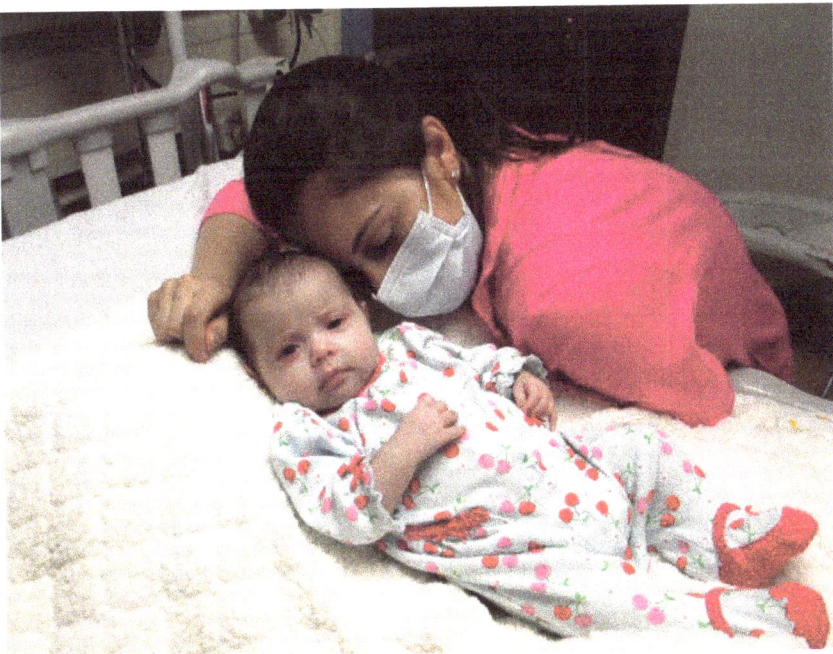

Evangelina and her Mom (family photo).

---

[1]https://blog.cirm.ca.gov/2014/11/18/ucla-team-cures-infants-of-often-fatal-bubble-baby-disease-by-inserting-gene-in-their-stem-cells-sickle-cell-disease-is-next-target/

Outside, for the first time ... (newsroom.ucla.edu).

Evangelina had to live in a plastic bubble all day and all night, or she would die. She had Severe Combined Immune Deficiency (SCID) which meant her immune system did not work. All the germs in the air that you and I don't think about? Those were deadly dangers to Evangelina. A simple cold could kill her.

If her mom or dad wanted to hug their little girl, they had to put on a nurse's mask. And when they left, Angelina stayed behind. Alone. If she was to live, she could never leave that little plastic room — except to go to the hospital.

And then one day, everything changed. Her mom and dad took Evangelina outside, and held her high in the air. She felt the sunlight for the first time. She opened her mouth, and laughed for sheer joy, happy a baby as ever was born.

And from that day to this, that baby and that family have lived a normal life.

What happened? Evangelina had been given a gift from the people of California.

Something small. Something wonderful. A California cure.

The first part of this story was told in my earlier book, *Stem Cell Battles: Proposition 71 and Beyond: How Ordinary People Can Fight Back Against the Crushing Burden of Chronic Disease: with a Posthumous Foreword by Christopher Reeve.*

P.S. That's Evangelina on the cover of this book.

1 THE ABSOLUTE MINIMUM YOU NEED TO KNOW FIRST

Let me tell you a story.

During World War One, the great entertainer Al Jolson volunteered to sing at a war relief performance. Famous for his renditions of "Mammy," "Swanee," "California Here I Come," and many more, Jolson was a kind and generous man, always glad to help a good cause. All he asked, and this he insisted upon, was that he would sing last, in the star's position.

But this time, the organizers tried to talk him out of it, saying he might want to reconsider, when he knew who the other singer was. Jolson refused, saying it was out of the question, no matter who it was.

The great night came at last, and the first singer was Enrico Caruso, the greatest opera singer of all time.

At the peak of his form, Caruso sang and sang: Ave Maria, La Done I Mobile, Vesta La Giubbia, songs he personally had made classics. By the power of his voice, he blasted away all troubles and worries, and gave the audience joy. They wept, and clapped their hands sore, refusing to let him quit, bringing him back again and again, song after song after song. And finally, when exhaustion stopped him at last, he bowed and left the stage, exiting to a thunderous ovation.

How could anyone follow such a performance?

But when the room went silent at last, Al Jolson bounded out onto the stage, beaming like the sun.

And he spread his arms and said:

"Hang onto your hats, folks, you ain't seen nothin' yet!"[1]

Hold that thought, please.

---

[1] http://www.jolson.org/nyt/nyt001022.html

"You ain't seen nothin' yet!" Al Jolson (Wikipedia).

Bob Klein, founder of CIRM, ICOC member Claire Pomeroy, Governor Arnold Schwarzenegger, author Don C. Reed, advocate Roman Reed (CIRM photo).

In 2004, California worked a miracle. Bob Klein, father of a diabetic son, raised $34 million for a citizens' initiative, Proposition 71, to build a state stem cell program. I had the good fortune to work beside Bob, first as a volunteer, and later as Vice President for Public Policy for Americans for Cures Foundation.

Proposition 71's goal? To challenge chronic disease and disability with $3 billion (with a "b"!) dollars' worth of stem cell research.

When we began, one scientist suggested lowering our expectations: with the slow pace of testing for federal approval, and the sheer difficulty of chronic diseases, (one definition for chronic is "incurable"), we would be lucky to bring even one therapy all the way to human trials. Thankfully, he was mistaken.

Today, the California stem cell program has not one but 42 therapies which have been tested, are being tested, or will be tested in human trials within six months.

We have begun to win. In addition to the "bubble baby" cure (more on Evangelina later), California has made tremendous progress in:

Paralysis: an embryonic stem cell therapy has already returned hand and arm function to six paralyzed patients; each of the people in the most-recent trial.[2]

Diabetes: a stem cell "credit card" can be slipped under the skin, where it will develop and provide the insulin the body needs.[3]

Blindness: people who were profoundly blind have regained a measure of vision, and one woman saw her own children for the first time.[4]

Scientists are using regenerative medicine to lead the charge against cancer, Alzheimer's Disease, arthritis, ALS, heart disease, HIV-AIDS, Parkinson's Disease, and more.

With continued funding, California will save lives, ease suffering, and make the world a healthier place.

In the words of the immortal Al Jolson:

"Hang on to your hats, folks, you ain't seen nothin' yet!"

---

[2] https://blog.cirm.ca.gov/2017/01/25/good-news-from-asterias-cirm-funded-spinal-cord-injury-trial/
[3] https://www.cirm.ca.gov/our-progress/disease-information/diabetes-fact-sheet
[4] https://blog.cirm.ca.gov/2017/03/21/a-stem-cell-clinical-trial-for-blindness-watch-rosies-story/

2 TO BREATHE, OR NOT TO BREATHE

In 1951, I was six years old and living in a hospital bed. I had asthma/bronchitis, which may not sound like much.[1] But, every inhalation was a separate wheezing effort, and a choice: to struggle and breathe, or relax and die.

There were relatives present: dad, mom, aunts/uncles/fidgety cousins; they were told I was on my way out, and to come say goodbye, take part in the event.

But the drama went on too long. My chest would stop rising, they would think, oh, that's it, surely he's gone this time, and then I would take another shallow breath.

At one point I imagined myself flying up to the ceiling and looking back, watching my family as they watched me: a skinny kid barely bulging the covers …

When I woke from the semi-dream state, a lady was sitting at the foot of the bed, knitting. I had no idea who she was. She was just there, and on the floor beside her was a cardboard box. She saw I was awake, smiled, stopped clicking her needles, and reached down into the box.

She took out a TARZAN comic book. On the cover, Tarzan was drinking with one hand from a forest pool. A deer and her fawn were watching him.

"You should read this," said the lady, "It's a good story. And it will be right down here on the floor, in this box, waiting for you."

---

[1]http://www.healthline.com/health/asthma/asthma-bronchitis#Overview1

Brigitte Gomperts (Newsroom:UCLA.edu).

Motivated by the comics, I got up soon, and the long recuperation began. There were injections of adrenaline which burned like liquid fire, and experimental drugs with long Latin names. Exercise stretched my ribcage and opened the breathing tubes; but it was years before the condition loosed its claws from out my chest.

What a vile thing is a lung disease! The wheezing sound of your own breathing keeps you awake at night. And since every activity requires oxygen, if you cannot get enough, you just stop. Even walking can be too much; I remember stopping on the way to the store, bending over, hands on knees, straining for air.

And lung disease leaves scars ...

Now, fast-forward sixty years. Enter Dr. Brigitte Gomperts, MD. Her studies focus on the repair and construction of the lung. Partially funded by California's stem cell agency, Dr. Gomperts works at UCLA, in partnership with the Eli and Edythe Broad Center of Regenerative Medicine.

Remember those lung scars? Dr. Gomperts is working to develop a cure for a deadly lung disease called Idiopathic Pulmonary Fibrosis (IPF). When IPF scars grow too thick, you cannot breathe, and the body shuts down.

"IPF is thought to affect more than 200,000 people in the USA ... mortality is very high. About 2/3 of patients die within five years of diagnosis ... IPF is on the rise and expected to double as the U.S. population continues to age ..."[2]

The only effective therapy? Transplantation.[3] Lungs are taken from an organ donor, and put into the chest of someone who needs them.

But suitable lungs are hard to come by, and the transplant procedure is expensive, roughly half a million dollars. Never 100% safe, lung transplants may bring infections: or the body may reject the new lung.

One problem is the lack of a good disease "model," a way to study the condition from beginning to end. If we can inject disease into a lab rat, we might find "markers" of the first signs of illness, and stop it early. But if we cannot find the beginning, it is hard to bring it to an end.

Also, animals may not have the same symptoms as we do. For instance, IPF lung scars on a mouse will go away after a while; those on humans never will.

But what if those diseased cells could be grown in a Petri dish...?

Using small chunks of skin from an IPF patient, Dr. Gomperts' lab made cells, with scarring very similar to the disease. Now she could follow its progress: actually watch the scars develop, and with no patients suffering.

Her team took blood samples from 250 patients with IPF, and converted them into cell lines to study.

Could stem cells heal a damaged lung, converting scars into healthy tissue? That may be one way to bring relief.

Also, Dr. Gomperts is working with bioengineers to build an actual working lung, to transplant.

Want the "recipe?"

First, ingredients. Mix three different kinds of lung cells: epithelial (surface of the lung), fibroblasts (structural support) and endothelial cells (lining for the tiny veins, the capillaries) and apply to a biodegradable structure, to give it shape.

---

[2] https://www.cirm.ca.gov/our-progress/awards/using-human-induced-pluripotent-stem-cells-improve-our-understanding-idiopathic
[3] https://www.ncbi.nlm.nih.gov/pubmed/22006881

Wait for it to grow into alveoli, "the tiny breathing sacks of the lung"…[4]

When those cells assemble, transplant as needed!

The replacement lung is smaller than the original, and perhaps not as efficient. But it might keep an elderly patient alive while waiting for a donor lung. Or, it could save the life of a child. A premature baby may need a new lung; as might a child with underdeveloped lungs, or one who is lacking sufficient breathing cells.

Try this: the next time you need to take a breath (i.e. now) … consider what will happen if you cannot complete the process.

To breathe or not to breathe, that is the question. It might not be Shakespeare, but it is life itself.

---

[4]https://stemcell.ucla.edu/research-focus

3 THE STRONGEST MAN IN THE WORLD

In the late 1960's, Olympic-style weightlifting was my life. I trained at Pennsylvania's York Barbell Club, where modern lifting essentially began, and became Associate Editor for their magazine, *Strength and Health*. In those days, everybody who was anybody in lifting came to York, and I had the privilege of meeting many of our sport's amazing champions.

One of America's greatest lifters was Steve Stanko. In 1941 Steve became the first man in the world to total 1,000 pounds in the three-lift combination of the military press (310), two-hand snatch (310) and clean and jerk (380).[1]

A vascular (circulatory) condition called phlebitis ended Stanko's competitive lifting. He never complained, but I saw him in the locker room once, and had to gasp. His legs were covered with dark blue lines, pencil-thick veins and arteries.

I asked him if it hurt, and he just laughed:

"HURT LAK HELL!"

For the iron-willed Stanko, vascular disease meant changing from Olympic lifting to bodybuilding — in which he also excelled, becoming Mr. Universe — but for others, more terrible consequences await.

One such condition is Critical Limb Ischemia (CLI), hardening of the arteries.

"... CLI ... may be present in as many as ... two million Americans ... it (may) result in amputation due to wounds that refuse to heal ...."[2]

---

[1] https://mastermuscle.wordpress.com/2012/09/14/steve-stanko-a-warriors-character/
[2] https://www.cirm.ca.gov/our-progress/arterial-limb-disease-fact-sheet

Steve Stanko (Strength and Health Magazine, March, 1942).

Right now, surgery is the main defense. Angioplasty (balloons pushed into the arteries) may nudge aside the blockage in the "hardened" arteries. The blockage may have to be cut with tiny knives, or frozen with helium. Stents may be used to tunnel around the blockage.

But nearly half of such operations are unsuccessful. Amputation may still be required — imagine your leg sawed off — and even that may not save the patient.

With the aging of our society, and the increasing numbers of diabetics at risk of limb amputations, CLI is an all too common threat.

Could there be a stem cell weapon? Spain and California hope to find out.

John Laird and Jan Nolta of the University of California at Davis are cooperating with Immaculada Herrera of the Hospital Universitario Reina Sofia in Spain.

How will they fight this terrifying condition?

Jan Nolta, Geralyn Annett, Teresa Tempkin, and Vicki Wheelock (ucdmc.ucdavis.edu).

A growth factor (VEGF, Vascular Endothelial Growth Factor) may help the body grow new arteries.[3]

VEGF alone will not do the job; it may not go where it is needed.

But what if we had a microscopic "emergency vehicle," to seek out the problem?

Mesenchymal stem cells (MSCs) can do that. Put into the body, MSCs travel to the trouble: the blockages in arteries and veins.

Add growth factor (VEGF) to the mesenchymal stem cell (MSC) and you have MSC/VEGF — the "paramedic van", self-directing and full of good medicine!

---

[3]https://www.cirm.ca.gov/our-progress/awards/phase-i-study-im-injection-vegf-producing-msc-treatment-critical-limb-ischemia-0

As the scientists put it, in their progress report:

"Mesenchymal stem cells ... are remarkably effective delivery vehicles, moving robustly through the tissue, infusing therapeutic molecules into damaged cells."

"... Injections of MSC ... have rapidly restored blood flow to the limbs of rodents who had zero circulation in one leg."

"We propose to use these MSCs as "nature's own paramedic system", arming them with VEGF to enhance ... blood vessel growth."[4]

Scientists from California have visited the lab in Spain, and vice-versa, making sure the joint effort will be the same, when the human trials begin. The only difference will be how the MSC/VEGF will be administered: Spain is using an injection into the blood; UC Davis will give the shot into the muscle.

Multiple safety and effectiveness experiments have been carried out, with each side bearing its own costs. If all goes well, the scientists intend to complete pre-clinical studies, move toward (FDA) regulatory approval, and initiate the clinical trials.

How are they doing? As I see it, they have advanced their methodology, bringing us closer to relief.

At the close of the 3rd (and final) year of their grant: "We have successfully engineered human MSCs to produce (high levels of) VEGF... and received ... approval to proceed to the FDA ... for the proposed stem cell gene therapy tria l..."

As Dr. Nolta put it, "None of this would have happened without CIRM. (CIRM is the California Institute for Regenerative Medicine, our state stem cell program, begun by Proposition 71.)" Thanks in large part to ... the California taxpayers who voted for Prop 71, stem cell therapies are changing the way medicine is done ...

"In the future, when the mysteries of stem cell therapies are more fully understood, patients may not have to endure such ... barbaric surgical techniques as limb amputation." — Jan Nolta, personal communication.

Steve Stanko would have been proud.

---

[4]https://www.cirm.ca.gov/our-progress/awards/phase-i-study-im-injection-vegf-producing-msc-treatment-critical-limb-ischemia-0

4 WHEN THE DOLPHIN BROKE MY EAR

Sleek gray lengths gliding through the blue, the dolphins were absolute masters of their environment …

From 1972–86, I worked as a professional scuba diver for Marine World/Africa USA, an aquarium-zoo in Redwood City, California. Most of the work was plain labor, scrubbing algae off the walls, floors, and windows of the giant tanks. But the creatures we swam with? Sharks, dolphins, eels, seals and killer whales — magic.

Sometimes we had to bring the dolphins to the vets for medical attention. This was not always easy. (In later years, tanks were made with rising floors, which made the situation less stressful for all concerned.)

But in those early days, we would often put a net in the water: not to catch the dolphins, but to narrow their swimming area. We would swim over the top of the net, wait until the dolphin appeared to relax, and then gently approach, and put our arms around them. When they cooperated, everything was easy. However …

One day a 350-pound dolphin named Ernestine snapped her jaws at me, and nodded her head in warning. ("Yes" means "no" in dolphin.) I should have backed off and tried again later. But I was young. I swam toward her, slowly, arms out.

Ernestine did a sudden forward roll, like dolphin judo. Her tail slammed the side of my head, impact like a car crash. Cold water entered my burst eardrum.

There was a roaring in my head, as I pulled myself out of the tank. Everything was spinning round and round, and I sat down to make the planet hold still.

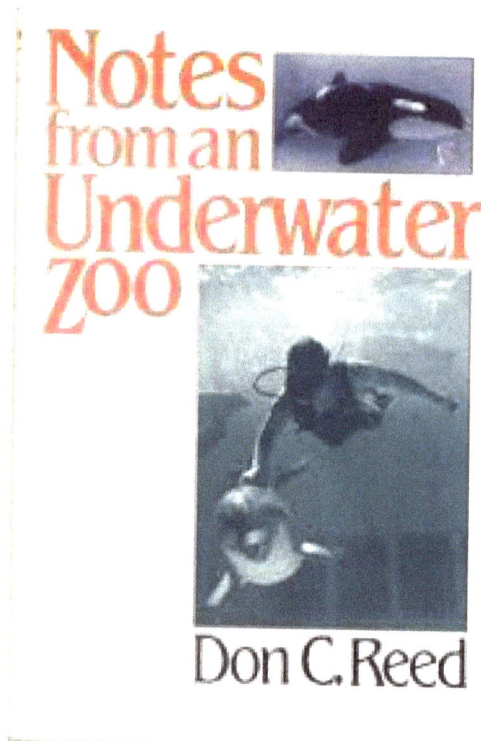

Author and dolphin.

I was lucky; the damage was minimal. I stayed out of the water for the next six weeks, and was deaf in one ear for a while.

Most people lose their hearing in a more permanent way.

Hair loss. No, not the white tufts sticking out of old folks' ears, but near-invisible hair cells, deep within the ear, in the snail-shaped cochlea, where 30,000 hair cells rest in a bath of liquid. When noise occurs, the hair cells tremble slightly. This sends vibrations to the brain, which translates them as sound.

When the hair cells wear out, there are no replacements. Lose some, and you become hard of hearing. Lose them all, and you are deaf.

Over the age of 65, one third of the world's population is deaf.[1]

---

[1]http://www.who.int/pbd/deafness/news/GE_65years.pdf

Alan Cheng (Stanford Profile).

And those who are just "hard of hearing?"

"Approximately 15% of American adults (37.5 million!) aged 18 or over report some hearing loss."[2]

Hearing aids help, turning up the volume, but in a distorted way. With or without the aids, hearing-challenged people face difficulties.

Anyone who has had to answer a hearing-challenged person's endless questions — "What did you say? What was that?" — knows how irritating it can be.

But imagine being on the other side of the equation, needing others to speak louder, not once or twice, but for every encounter of the day? And what if the hearing departs altogether?

Helen Keller, blind and deaf, compared the two conditions:

"Deafness means the loss of the most vital stimulus … voice that brings language, sets thoughts astir, and keeps us in the company of man. Blindness separates people from things; **deafness separates people from people.**"[3]

---

[2] https://www.nidcd.nih.gov/health/statistics/quick-statistics-hearing
[3] http://www.goodreads.com/quotes/391727-blindness-separates-people-from-things-deafness-separates-people-from-people

All too often a deaf person will sink into loneliness, depression and despair, and just quit trying to participate. And, of course, some jobs are denied them.

But if their hair cells could regrow, we might restore their hearing.

Why are there no deaf chickens? Birds can regrow hair cells. If they can do it, why should humanity be denied the subtlety and grandeur of sound?

At Stanford University, a group of scientists want to rectify that situation.

"There are seven of us in the Stanford Initiative to Cure Hearing Loss (SICHL, pronounced like sea shell) working on different parts of the ear. As a group, we collaborate, and push hearing research forward," said Dr. Alan Cheng, M.D., when we did a recent catch-up interview.

A surgeon, Dr. Cheng operates on deaf children, giving them a cochlear implant, a "complex electronic device that can help provide a sense of sound ..."[4]

But implants give only a crude imitation of normal sound. The person may understand speech, and grasp warning signals, like a car honk, but that is about it.

Dr. Cheng feels frustrated, because he does not have something better for the children. But that may change.

Remember how I felt dizzy after the dolphin struck? Not only was my hearing affected, but also my sense of balance. Our sound system has two parts: hearing itself, and also balance. The hair cells in the hearing part do not regrow, but those in the balance system do.

A tiny balance organ, the utricle, is our center of gravity awareness. When the elevator drops, or a pilot swerves the jet, what we feel are signals from the utricle.

And inside the utricle, hair cells can regrow.

Dr. Cheng may have found the stem cell that starts hair cell regeneration.

"How's it going?" I asked him.

---

[4]https://www.nidcd.nih.gov/health/cochlear-implants#a

"It's going!" he said enthusiastically, "We see re-growth of hair cells in the mouse balance organs. And the balance function appears to improve, according to how many hair cells come back."

And then, an unexpected delight, like money found between couch cushions.

There may be a way to <u>prevent</u> a major cause of deafness.

In poor countries, sick children are often given an antibiotic drug called aminoglycoside. This helps their immune system problems — but may also make them deaf. An estimated 30% of those who take this drug lose their hearing.[5]

What if there was a replacement antibiotic that would not steal the patient's hearing? Dr. Cheng and colleagues may have developed a drug to do just that.

Their path is complicated and difficult. In fact, one of their group leaders, the renowned Dr. Stefan Heller, has just released a paper on deafness-healing approaches that did NOT work out for them.

Discouraging? Not really. The scientists at SICHL — and all who read the paper — can use that knowledge to avoid mistakes, not wasting their time going down the wrong road. And it offers a positive direction: on the importance of the environment of the inner ear.

Dr. Cheng sums up the struggle:

"Regenerating hair cells in humans to restore hearing is going to be a long journey with numerous obstacles. But with the support of the California Institute of Regenerative Medicine, and the Stanford Initiative in Curing Hearing Loss ... we are ... positioned to overcome these challenges in years to come." — Alan Cheng, personal communication.

Scientists like Drs. Cheng and Heller are self-starters. If they have the funds, they will do the work: practically living at the lab, working toward the day when the deaf can hear again.

If I had to sum up their attitude, it might be something like:

If we never quit trying, we can only win or die — and everybody dies, so why not try?

---

[5]http://emedicine.medscape.com/article/857679-overview

5 THE BOY WITH BUTTERFLY SKIN

One day in the eleventh grade, I did a thousand sit-ups without stopping; I don't know why. I just started doing them in PE class, going for a hundred at first, and then continuing on and on. The person holding my ankles got bored and quit; somebody else tossed a cupful of water on the floor where my back kept touching. But I was on a mission and kept on and on, a skinny teenager doing sit-ups …

I wore a patch of skin the size of a silver dollar off my tailbone area. It was raw and itched for a couple of days, and then healed without complication.

But what if the skin could not heal?

Try this. Take your right index finger and run it roughly across your left forearm. Nothing happens, right? You see the skin ripple, but it springs back like before.

If you had a skin condition called Epidermolysis Bullosa (EB), that small touch could leave a blister, or expose raw flesh.

When you first meet John Hudson Dilgen, in the video "Boy with Butterfly Skin,"[1] he appears to be wearing white sweats. He is smiling and talking, a regular kid, the athletic type who would be running around like crazy at the school playground, and be last to come in from recess.

It takes a moment to realize the "white sweats" are bandages over wounds.

The title "Butterfly Skin" comes because EB skin is fragile as a butterfly's wing.

---

[1]https://www.youtube.com/watch?v=oM5Am04ellE

John Dilgen (www.johnhudsondilgen.com).

In a healthy body, the layers of skin stick together by the body's natural glue, collagen. In EB, the collagen gene does not work right, and the outer layer of skin can easily break away.

When John was born, he had no skin on his feet from just the strain of childbirth. As he grew, more terrors developed.

"Bathtime," said his mother, was "heartbreaking, with relentless fear and screaming ..."

He loved potato chips, but just swallowing something rough can be deadly for a child with EB, doing damage to the inside of the throat.

And the parents? Imagine being afraid to hug your child, lest you damage them!

Today John Dilgen is 13, but the disease is still with him. If he rubs his eyes, the damage may require him to spend several days in a completely dark room.

EB is rare, an "orphan disease." This is good news/bad news: we want it rare, because no child should have to suffer like that, but a small

Anthony Oro (Stanford Health Care).

number of patients also means fewer customers, less profit for the big corporations.

But there are practical reasons to support "orphan disease" research. Other conditions are similar: improvements for one may help another.

EB is rare, but skin disease is common, affecting 2% of the population. Wound healing in general affects millions. Finally, research on EB may accelerate cure for cancer. EB sufferers often die of a deadly cancer called squamous cell carcinoma. Studying the skin of an EB patient may determine the trigger-point of many kinds of cancer — exactly when and how it begins — potentially benefiting millions.

"... treating rare diseases is the first step toward achieving ... (the) personalized medicine that the Obama Administration has highlighted as a priority." — Julia Jenkins, Executive Director, Every Life Foundation for Rare Diseases, personal communication.

Stanford University is heavily invested in this effort, with a superb dermatology lab.[2]

Here are three Stanford scientists, working on the problem right now.

They are: Alfred Lane, Primary Investigator (PI), an expert at FDA regulatory affairs and former Director of the EB clinic at Stanford; Marius

---

[2]http://med.stanford.edu/dermatology.html

Wernig, stem cell reprogramming authority, who once came to a formal science event wearing white tennis shoes with his tuxedo; and Anthony Oro, a skin genetics and development expert, who is not only attempting to do the impossible with the invisible in medical research, but he can even explain it in people-talk.

How will they try to defeat this presently incurable disease?

One weapon they may use is Induced Pluripotent Stem cells (iPS), the micro-manipulation of skin cells which won Shinya Yamanaka the Nobel Prize.

And if they win? "There are over 200 genetic skin diseases that could in principle be treated with this approach. Similar therapeutic reprogramming studies are being developed for other tissues like heart, brain, pancreas, liver and cornea. The 'road plowed with skin' can make the manufacturing and regulatory tracks much easier for subsequent therapies." — Anthony Oro, Stanford, personal communication.

Here is my layman's interpretation of what they plan to do:

1. Take a tiny skin sample from the patient;
2. Using Professor Yamanaka's iPS process, make personalized stem cells from the patient, "regressing" the cells back to an embryonic-like state;
3. Correct the defective genes;
4. Do quality control on the corrected iPS lines to remove those with genetic abnormalities;
5. Make new cells (with the corrected genes);
6. Make the healthy new skin and graft it onto the patient's wounds.

Have they won? Not yet.

One attempt has failed; another is intended. Like a series of advancing waves, science moves forward, every effort benefiting from what went before.

Will the Lane/Wernig/Oro approach be the one that succeeds? We cannot know. But the fact that these three talented individuals are fighting on gives me hope.

"Stanford ... will soon move forward with an expanded patient study ... We are hopefully years, not decades, away from meaningful

Jason Reed (family photo).

therapies and potential cures." — Jessica Schneer, Epidermolysis Bullosa Research Partnership Executive Director, personal communication.

My grandson Jason plays baseball, and does not worry about his skin; every kid deserves such fun and ease of mind.

Bumped elbows and skinned knees should be a normal part of a child's life, a minor inconvenience, nothing more, and epidermolysis bullosa should be remembered only because it is a hard word to pronounce.[3]

Want to follow the research to fight "butterfly skin"? See number[4] below:

---

[3]https://ebresearch.org/what-is-eb/
[4]https://ebresearch.org/research/

6 THE GREAT BALDNESS "COMB-OVER" REPLACEMENT?

Some male members of my family might politely be described as "hair-challenged": they are not shiny-domed like the cartoon character Elmer Fudd, but in that ballpark. No one wears a rug; but some do the great comb-over, with one side allowed to grow long, and then flipped across the top in hopeful camouflage.

I first became aware of this problem about ten years ago, when the shower drain started clogging up with hair.

I told my barber Raja about it, and she gave me a secret. I should really not share it with you. If word gets out, something which is currently cheap and plentiful may become expensive. But I will tell you anyway, a page or two from now.

First, consider an astonishing concept.

Hair communicates.

Think of one single hair stalk. If you pluck it, or if the skin around it is damaged, that hair will give out a tiny chemical "scream": not a sound, of course, but a signal, from stem cells all over its surface. If enough such signals go out, the hair may shift into growth mode.

What if a bald person's hair follicles (the hair stalk holders) could be tricked into shifting into growth mode?

He or she might become hairy once more.

This could be more than just vanity; scientists like Dr. Maksim Plikus of UC Irvine study *hair as part of the healing system*. Remember those stem cells on the hair stalks? Some fight infection; others help bring together the edges of a wound.

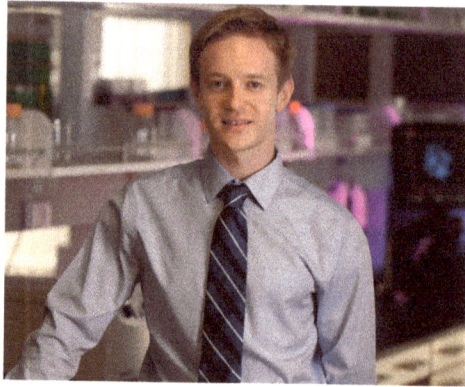

Maksim Plikus (devcell.bio.uci.edu).

"Our laboratory (seeks to) understand the natural limits of stem cell plasticity in response to injury ... In the center of large skin wounds, cells can acquire an embryonic-like state and develop new, fully functional hair follicles. Additional tissues regenerate ... (This) can be so efficient that several months after wounding, scar tissue can hardly be distinguished from the normal skin ..."[1]

It takes a lot of signals to restart hair growth. This is called quorum activity, like when a committee needs a certain number of people to make a vote count.

Ever watch birds getting ready to fly south for the winter? They gradually gather, and make short hopping flights here and there, nowhere in particular, just practicing, figuring out who fits where in the flock. Then one day there are enough of them, and conditions are just right. They burst into the sky, swirling up like ashes in the wind — and then, having patterned their quorum, the birds head south.

"Quorum activity" in the body is similarly crucial. In the intestine, for example, hair-like villi work in waves, sending microscopic signals to pass the poop along.[2]

---

[1] http://devcell.bio.uci.edu/faculty/maksim-plikus
[2] https://www.scientificamerican.com/article/hair-regrowth-discovery-suggests-skin-cells-communicate-like-bacteria/

Cheng Ming Chuong (USC Pressroom).

Plikus's mentor, Dr. Cheng Ming Chuong, organized one experiment, in which 200 hairs were pulled out of the backs of anesthetized rabbits.[3]

Two hundred hairs were removed; <u>twelve hundred hairs grew back</u>. The body apparently used hair growth to help heal an injury.

Interestingly, it only worked if the removed hairs were close to each other. That apparently was regarded as a wound to be dealt with. If the hairs were far apart, the plucking was ignored, a minor irritant.

A surprise: healthy cells grow strongest at different times of the day. To get a longer lasting shave, do it at night. Shaving in the morning brings quicker re-growth, the famous "five o'clock shadow."

For people undergoing cancer therapy, Cheng and Plikus' experiments may help.

Cancer cells and regular cells multiply at different rates, even varying by the time of day. If we know when the good cells are most quiet, that is when radiation should be done.

---

[3] https://www.ncbi.nih.gov/pmc/articles/PMC4393531/

We want to blast the bad cells, not the good ones. Zapping healthy cells too much may bring side effects: nausea, diarrhea, hair loss, other nasty stuff.

By altering the timing of the radiation, we hurt fewer good cells, and may lessen patient suffering.[4]

As a financial investment, the hair regeneration industry looks like a good bet.

"Japan's largest research organization Riken … has teamed up with … Kyocera and Organ Technologies, to develop a (baldness) cure based on regenerative medicine … targeting 2020 for commercialization. (Also) scientists at Sanford-Burnham Medical Research Institute in California … made a similar announcement citing similar methods."[5]

These are responsible organizations. Unfortunately, other, less scrupulous companies calling themselves stem cell clinics are offering (not cheaply) to regrow bald people's hair. It might work, it might not; and it could have major side effects.

Anything connected with regenerative medicine should have an FDA approval before being made available to the public.

But, we must not forget my colossal "secret?" Are you ready for this?

Aloe vera. The clear gel made from cactuses costs a couple bucks, and can be found on the skin care shelf of your local grocery store.

As technology goes, it is pretty basic. I put a dab of glop on my left palm, rub my hands together, then massage it deep into my scalp.

My proof? About three days after I started the aloe vera, the shower drain stopped clogging up. My hair loss apparently diminished.

Scientific? Not at all. Just the massage effect? Could be. But, I have been doing it for about ten years now. I stopped using it once, to see what would happen. The drain plugged up almost immediately. I went back to the aloe vera, and the shower drain is clear once more.

I do not believe aloe vera regrows hair — but it might help to keep some foliage — and it absolutely feels nice when you rub it in.

My hair is white, but it is still there. I intend to keep using that cheap cactus gel for the remainder of my years, or at least until it stops working.

---

[4] http://www.ocregister.com/articles/cells-520863-hair-radiation.html
[5] http://www.forbes.com/sites/jboyd/2016/07/13/stem-cells-to-make-hair-today-gone-tomorrow-a-thing-of-the-past/#6b11527a3cc5

7 "HE SEES! HE SEES!"

In Jules Verne's classic novel *Michael Strogoff*, the hero is falsely accused of treason by the villainous Ivan Ogareff. In a devastating scene, the Czar's executioner places a red-hot sword close to Michael's eyes, blinding him.

Strogoff stumbles his way through the story, and then, confronted by the mocking evil antagonist, he suddenly cries out for justice and challenges Ogareff to a duel. The villain is glad for the chance to put an end to Strogoff and comes at him disdainfully. But Michael duels him, blocking, lunging, fighting intelligently. "He sees, he sees!" shouts the villain in astonishment, and Michael kills him, which of course makes him innocent of all charges.

How did he do it? At the precise moment the heated metal had been put beside his eyes, Strogoff was thinking about his mother. Tears insulated his eyeballs just long enough, and he was not blinded.

For most blind people, of course, there are no miraculous rescues.

They live in darkness: reading by fingers on the raised-bump Braille system, walking with the investigatory white cane, perhaps assisted by one of the famous "Seeing Eye Dogs," if they live where such help is available/affordable.

For the blind, neatness is a necessity. They must know exactly where everything goes, or how can they find it again? I think of my messy room, a cheerful chaos of stacked books and binders; how would I cope without sight?

In addition to the loss of independence, blind folks are more likely to fall than sighted people, with risk of bone fractures and hospital costs.

"Michael Strogoff" (Classic Comic Books, Zvi Har'Els Jules Verne Collection).

But now, two scientists are dueling it out, not with each other, but against blindness itself, a condition which has plagued humanity since mankind began.

Mark Humayun and Henry Klassen: let's jump back and forth between them.[1]

Dr. Humayun is fighting the most common cause of blindness among people past 60: Age-related Macular Degeneration (AMD), a gradual but

---

[1] https://blog.cirm.ca.gov/2016/08/29/seeing-is-believing-how-some-scientists-including-two-funded-by-cirm-are-working-to-help-the-blind-see/

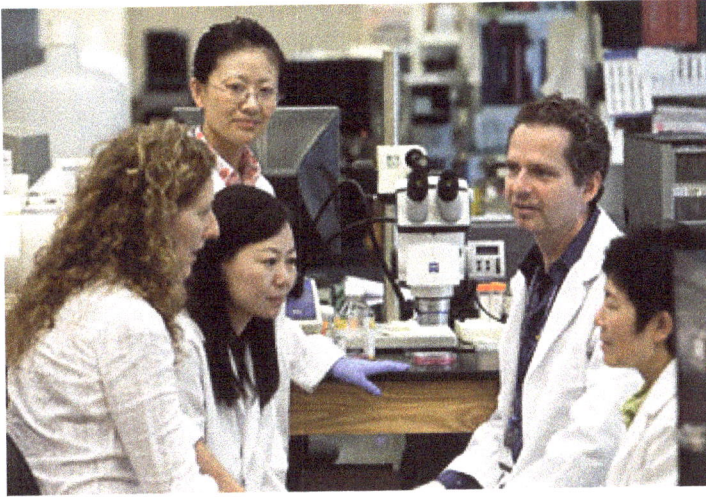

Henry Klassen (Blindness.org).

irreversible loss of sight. At first, just the center of your vision is lost; as if a big dot was in front of your eye. You can look around the edges for a while. But gradually the dot widens, big as your thumb, and finally, it expands too much, and you cannot see at all.

In the U.S., this affects roughly one in four people over the age of 65.

Dr. Klassen is taking on Retinitis Pigmentosa (RP), which steals vision from three million people worldwide.

Humayun is employing embryonic stem cells, those "blank check" cells which can be turned into any kind of body cell required.

At the University of Southern California (USC) Humayun's team is trying to restore the <u>support cells</u> at the back of the eye. These cells provide nutrition and protection to the light sensing cells (photoreceptors) of the retina.

At the University of California at Irvine, (UCI) Klassen's team aims to support and possibly restore the <u>photoreceptors</u> themselves. If successful, this may "provide a stepping stone to many otherwise incurable diseases of the brain and spinal cord."

Humayun's team delivers the cells on a tiny scaffold, put on the back of the eye.

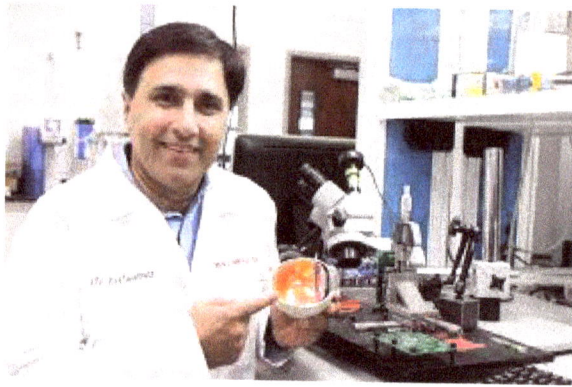

Mark Humayun (eye.keckmedicine.org).

Klassen injects the materials into the (numbed) eyeball itself.

How are they doing?

Humayun: "In this ongoing phase (of the) clinical trial we are assessing safety of the implant. The study will include two cohorts, groups of 10 patients ... We will assess the vision of patients to see if there are any changes in their ability to see."

Klassen: "Over the past year, significant progress has been made in advancing our therapy for retinitis pigmentosa (RP) ... Eight patients in our first cohort and one patient in our second have been injected at two different dose levels: 0.5 million and 1.0 million cells ... Our safety data (was) reviewed ... no safety concerns."[2]

One of these patients was Rosie Barrero: A beautiful young woman, Ms. Barrero does not give the appearance of being blind. Her eyes follow you as you speak, but she is guided by hearing, not sight.

As she puts it, "I cannot go outside the house by myself."

Rosie lost night vision first, then was called "near-sighted" by a school nurse, but then everything got worse, until finally the curtain came down. She became completely blind, just before she gave birth to twins.

---

[2] https://www.cirm.ca.gov/our-progress/awards/retinal-progenitor-cells-treatment-retinitis-pigmentosa

But today, having had just the minimal amounts of treatment given in a safety study, she has had a measure of vision restored. She can see outlines, and pale colors, and now ...

"I can see my own handwriting again," she says, almost in disbelief, "and most of all, I can see my twins ..."

Dr. Humayun? His test results are not as far along as Dr. Klassen's.

But we will not be impatient.

This is the man who was recently honored by President Obama, for a non-stem cell invention, a combination eye implant plus camera-glasses, which gives blind people a level of vision they did not have before.[3]

We are lucky to have such men involved: they and their partners, champions like UC Irvine's Jing Yang and USC/UCSB's David Hinton and Dennis Clegg.

And one more thing.

Remember Michael Strogoff and the fictional miracle of his not-lost sight?

Author Jules Verne, a supporter of science even in a historical novel, based the blinding scene on an actual phenomenon, the Leidenfrost effect, which has to do with the heat-insulating effects of liquid, so it was just barely possible Strogoff's vision might actually have been spared ...[4]

"He sees! He sees!"

What beautiful words.

---

[3] http://www.prnewswire.com/news-releases/president-obama-honors-usc-eye-institutes-dr-mark-humayun-with-national-medal-of-technology-and-innovation-300199517.html

[4] https://en.wikipedia.org/wiki/Leidenfrost_effect

8 COP AT THE WINDOW

Taptaptap …

Roman Reed twitched in his sleep, hunched over his van's steering wheel.

TAPTAPTAP … "Sir?"

"Hmmf, what?"

Roman blinked at the cop's flashlight's glare, shining through the window.

"May I ask what you are doing here?"

Roman ignored the obvious answer, (that he had been sleeping until someone woke him), and explained that he had been speaking at the Roman Reed Laboratory at the Reeve-Irvine Research Center in Southern California, and he had another engagement at 9:00 in San Francisco, and pulled off the road to catch 40 winks …

"Will you step out of the van, please?"

"I will," said Roman, "But it is inconvenient."

"Why?"

"That's my wheelchair back there."

The policeman turned his flashlight on the powerchair in the dark.

There was a pause. Then the cop's voice changed.

"You're paralyzed, and doing all that?"

As any spinal cord injury researcher can tell you, my son inspired the Roman Reed Spinal Cord Injury Research Act of 1999. From that act sprang fifteen million dollars of state funding: $15,126,100, plus $87

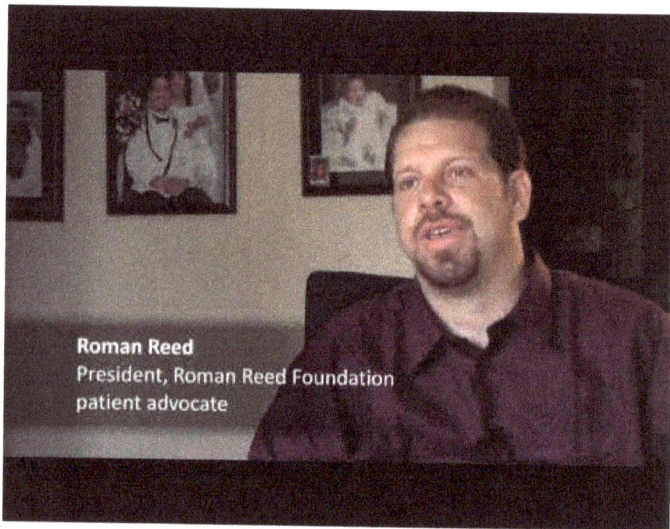

Roman Reed (amsvans.com).

million more in add-on grants from the National Institutes of Health and other sources.[1]

It was a "Roman's Law" grant which first funded Hans Keirstead's pioneering work on human Embryonic Stem Cells, (hESC) which went on to become the world's first hESC clinical trials, now underway at Asterias Biotherapeutics.[2]

But few realize what a non-stop act the Cal Berkeley graduate is. There is a reason he won the Inspiration Award from the World Stem Cell Summit, the Stem Cell Person of the Year Award from the Paul Knoepfler Lab, the Willie Shoemaker Award for Advancing Awareness of Spinal Cord Injury, and a bunch more.

If he was to describe himself, Roman would probably say he is the Founder and CEO of StemRemedium, a biotech company focused on pioneering stem cell research for neuroregeneration, but he is also the CEO of Sacred Cells Research Partners, LLC, has served as Chairperson

---

[1]http://www.reeve.uci.edu/roman-reed.html
[2]https://blog.cirm.ca.gov/tag/roman-reed/

on the Fremont City Planning Commission, President of the Roman Reed Foundation, was Executive Director of Community Relations for Stanford's Spinal Cord Injury Repair Committee, etc., etc. — while still making time to help coach his children's sports teams.

He doesn't stop.

I remember the terrible night of his football accident, September 10th, 1994. We were in the back of the emergency vehicle, on our way to the hospital, and Roman said:

"Am I ever going to squat 500 again?"

"I don't know, son," I said.

"I have to know," said Roman.

What was I to say? He was paralyzed from the shoulders down, a tetraplegic, unable even to close his fingers. He deserved an honest answer. But all I could think to say was:

"I would never bet against you."

Roman said "mmf," and that was that.

In the hospital he had a banner made, "I can, I will, I shall!" which was kept hanging over his bed.

It has been twenty-two years since the accident, and he has not wavered.

One doctor told him, "You are in denial," and Roman agreed with him. And to this day, he is still denying the incurability of paralysis.

From his own experience, he knows there is at least the possibility of recovery. With the help of the most advanced experimental therapy available (a drug called Sygen, then undergoing FDA clinical trials, which we obtained on our own) and vigorous rehab exercise, Roman recovered an unusual amount of triceps function, those important muscles on the backs of the arms. That might not sound like much, but it happened well past the time when recovery of any sort is considered possible and is extremely useful in living an active life.

Roman has taken on a clearly impossible challenge: to find a cure for paralysis. He intends to win, and every day he works to make it happen. From the moment he wakes in the morning, and does his email (which takes several hours) he is always looking for some new approach for research, power person to contact, or just a place to talk: to raise awareness.

He speaks to student researchers at Cal Berkeley, to scientists fighting for grants; and people he meets by accident, like the Hells Angels motorcycle rider, who slapped him on the shoulder and said, "Got your back, bro!"

When Roman hears of a person newly paralyzed, he always tries to contact him or her, to share with them the reasons for hope: including the clinical trials going on right now, with major funding from the California stem cell program.

Abraham Lincoln's law partner William Herndon once said of Lincoln that "his ambition was an engine that never slept,"[3] and in a way that is true of Roman.

He may have been sleeping by the side of the road, but his eyes are always open for another way to fight for cures.

---

[3]http://www.sites.si.edu/exhibitions/lincolnSample.pdf

9 "GO WEST, YOUNG (WO)MAN" — TO A BIOMED CAREER?

Want to save lives, ease suffering, help your country, and make a very good living as a biomed scientist or laboratory technician?

If so, California is the place you ought to be.

Not that there is anything wrong with other great science states, like Connecticut, New York, Massachusetts, Maryland, Delaware, or New Jersey (to name a few).

But California is the unquestioned biomed leader of America and the world.[1]

Might there something here for you? Check it out; it might change your life.

Right now, California's life sciences (biomed) employs 287,200 men and women. That is more jobs than the entire state's Internet business (192,069) or the motion picture industry (163,907).

When you add the indirect jobs (to fill needs created by the biomed companies) — another 597,000, that's 884,200 people working in the biomed sector — it totals to nearly a million.

Average wage? $116,000 a year.[2]

Does California still need more biotech workers and researchers? In a listing of the 12 "hottest" regions in the country for biomed expansion, at

---

[1] http://www.chi.org/wp-content/uploads/2014/11/2015-CHI-PwC-California-Biomedical-Industry-Report_Final.pdf
[2] http://califesciences.org/2017report/

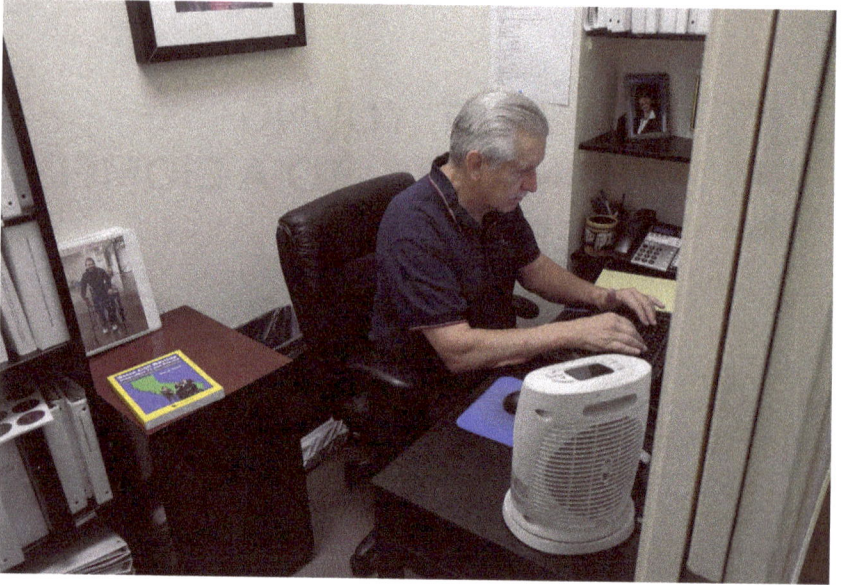

Don Reed happy (personal photo).

the top of the list was: Northern California. And number two? Southern California![3]

Which major cities have the greatest demand for new biomed engineers? Of the top eleven American cities, the most employee hungry were six (Los Angeles, San Diego, Oakland, Anaheim, San Francisco and San Jose) all in California.[4]

AND, the California stem cell program might even help your career.

CIRM (the California Institute for Regenerative Medicine) is working hard to help young people find jobs in biomedicine: three ways.

First: by developing a college course on biomedicine and stem cell research, and making it available (free) to every California high school or college.

---

[3] http://www.biospace.com/News/top-12-hot-biopharma-regions-for-growth-and/347389

[4] http://www.insidermonkey.com/blog/11-cities-with-the-highest-demand-for-biomedical-engineers-399379/?singlepage=1

Senator Art Torres encourages minority students to investigate a biomed career (cirm.ca.gov).

This is the result of a law, SB 471, (Romero, Steinberg, Torlakson), the California Stem Cell and Biotechnology Education and Workforce Development Act of 2009. The law was made because biomedicine is growing so fast there may not be enough trained people to fill all the coming jobs.[5]

The curriculum is challenging but doable.[6]

Want a bunch of free stuff to help you get ready for this great new field, like videos, groups to join, and other resources?[7]

Second: if you are a High School student, or you know someone who is, would you like a summer job working in a stem cell lab? It might be possible.

SPARK, (the Summer Program to Advance medical Research Knowledge), hires high school students to work in accredited laboratories doing actual science — projects they choose — under the guidance of

---

[5] http://tinyurl.com/hsjzgt8
[6] https://www.cirm.ca.gov/our-impact/education/stem-cell-portal
[7] https://www.cirm.ca.gov/our-progress/student-resources

professional scientists. Here are some of the students, telling you what it was like.[8]

Want to apply? For the application form, go here:[9]

And third: Maybe you are in college already, working on your Bachelor's or Master's degree? Would you like a salary, to learn about stem cell research?

CIRM's "BRIDGES" program gives college students a paid job at a nationally-renowned laboratory.

Its goal? CIRM's then-president, Randy Mills:

"The goal of the Bridges program is to prepare undergraduate and Master's level students in California for a successful career in stem cell research. That's not just a matter of giving them money, but also giving them good mentors who can help train and guide them, of giving them meaningful engagement with patients and patient advocates, so they can have a clear vision of how the work they are doing can impact people's lives."

Does it work? Jonathan Thomas, Chairman, California stem cell Governing Board:

"The Bridges program has been incredibly effective in giving young people, often from disadvantaged backgrounds, a shot at a career in science. Of the more than 700 students who have completed the program, 95 per cent are now working in a lab, enrolled in school, or applying to graduate school in biology. Without the Bridges program, this kind of career might have been out of reach for many of these students."[10]

These are students of diverse backgrounds, many of whom would not have gotten that far without the financial assistance of the Bridges program.

Would biomed be a good fit for you? Only you can know, or find out.

---

[8] https://blog.cirm.ca.gov/category/events/cirm-spark-program/
[9] https://blog.cirm.ca.gov/category/events/cirm-spark-program/
[10] https://blog.cirm.ca.gov/category/events/cirm-creativity-program/

If you are in high school or college now, talk to your guidance counselor, ask about an introductory course in biomed or stem cells, and see what your school has to offer.

If a career in life science sounds interesting, you owe it to yourself to check it out.

To work hard all day in an effort to save lives and ease suffering?

What could be a nobler career?

10 AND HOW WILL YOU BE PAYING FOR THAT NEW HEART?

My greatest fear is the loss of Gloria, wife and best friend of 48 years.

Every day begins and ends with her. I get up around 1:30 AM, so I can write without being disturbed; she wakes up two or three hours later. When she clears her throat a certain way, I go climb in bed and we snuggle up. We talk and laugh for a half hour, and then get up and go to the gym.

At night we play a game of cards, which she almost always wins, having the ability not only to remember which cards I pick up, but also to use mental telepathy on me. In her mind she will tell me to put down a three of clubs or whatever, and I will have a nice warm feeling that I really should put down the three of clubs …

But all things end, and one day life will part us. We cannot know who will go first, of course, but Gloria's family has a history of heart failure, and she is a tad overweight. Odds are she will leave this world by America's most common cause of death: heart attack, which kills 1/3 of us.

But to never hold her hand again? To call out "Hon!" when I get home, and not hear her cheerful response? I cannot accept that, not until we are a hundred.

So, we fight the problem! We try not to eat too much (hard because she is a GREAT cook), and exercise — at least a little — every day.

But a couple nights ago, watching TV before going to bed, Gloria told me she felt that pressure on her chest again, right now.

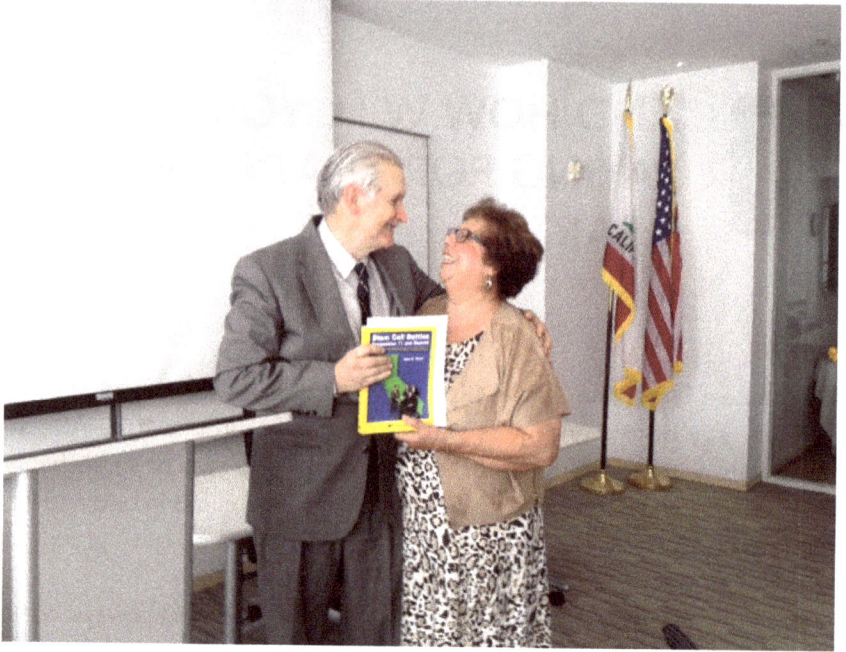

Don and Gloria Reed (cirm.ca.gov).

I said "Oh!" and covered my face for a second, because every time this happens, I think it is the end. But hiding does not help, and presently I jumped up and ran for the baby aspirin. She took two. In five minutes the pressure went away, and everything was back to normal. (In her purse is also a container of nitro glycerine pills, to put under her tongue in case of a severe attack.)

If there was a way to make her heart young and strong, of course I would jump at it — if we could afford it.

But what if the cure for heart disease was a million dollars?

I would have to watch my wife slip closer and closer to death, while the rich take out their credit cards, and have their hearts repaired.

It's like when that (censored) Martin Shkreli jacked up the price of a life-death important medicine from $13.50 for a daily dose to $750. Can you imagine that, to raise the price of one pill from thirteen bucks to seven hundred fifty dollars? There is no excuse for such greed. Shkreli wasn't even ashamed of it, saying: "Everybody's doing it..."[1]

So how should pricing be determined for a drug or therapy developed on a CIRM-funded grant?

Not an easy question. For a new cure to be available, corporations must manufacture and sell it. If there is no profit motive, would they invest the billions required to develop the product?

CIRM held many long public meetings on this subject: a bitter disagreement, dividing scientists, businessfolk, legislators and patient advocates. It became unbelievably complicated; here is an ultra-short version.

There appeared to be three positions in the argument:

One: Any medication which came from a state-funded grant should be made available to Californians for free;

Two: The cure itself was reward enough; let the market charge what it wanted;

Three: A compromise: for low and middle-income folks, a discount.

The answer (when it came) seems logical now, but it took a ton of wrangling to get there.

Everyone who takes a grant from CIRM must agree on certain stipulations.

First, they must share the knowledge developed, so their "research findings can be replicated by others."[2]

Second, if there is a "Bombshell" breakthrough, "a direct monetary payment" must be given to the state stem cell research program: the amount varying on how much CIRM contributed.

Third and most important, any marketable product developed with CIRM funding must have an accessibility program. The scientist or corporation must develop this program, to be approved by the ICOC before being released to the market.

---

[1] https://www.bloomberg.com/news/articles/2016-12-23/martin-shkreli-says-of-course-he-d-raise-drug-price-again
[2] https://www.cirm.ca.gov/sites/default/files/files/agenda/101106_item_11.pdf

Senate Bill 1565 (Dymally) put the CIRM-developed policy into law:

"... the for-profit grantee agrees to provide a plan...to provide uninsured California residents access to resultant therapies... Also, grantees will provide the therapies at a discount price to entities that purchase them in California with public funds."[3]

So discounts will be made available for low and middle income people like us.

For a clarifying series of questions and answers on the "Intellectual Property and Revenue Sharing Requirements," go to the source itself: CIRM.[4]

P.S. Do you want to know the secret of a happy marriage? <u>Love is</u> a verb, an action word, an activity; not something you fall into or out of. If you work at it every day, you can make love last.

Example: every time Gloria cooks a meal for me, I say "Thank you," or make some small compliment. I never want to take her for granted. Because as soon as I think, okay, we don't have to work at the marriage any more, that means we are just about to have a fight.

Love is a verb, and it only took me 48 years to figure that out!

---

[3] ftp://www.leginfo.ca.gov/pub/07-08/bill/sen/sb_1551-1600/sb_1565_cfa_20080616_114123_asm_comm.html
[4] https://www.cirm.ca.gov/our-funding/chapter-6-intellectual-property-and-revenue-sharing-requirements-non-profit-and-profit

11 THE ANSWER TO CANCER?

December 30, 2016. I was sitting in the outer office of the Kaiser hospital lab, waiting for a blood test. I have one every four months now, to see if the cancer has come back.

It has been two years since the words, "You have cancer," echoed in my brain. As I sat waiting for my number to be called, highlights of the adventure returned.

All dignity was stripped away. Medical samples? Imagine lying nude face down on the exam table while something like an oversized toenail clipper goes up inside you, nipping out chunks of flesh. Technically, it does not "hurt" because they numb the area, but you still feel — and hear! — each click as the body is diminished.

When the results came out positive for the big C, I had to decide what to do.

Surgery? Radiation? Hormone therapy? I opted for all three.

First, they cut, removing the prostate gland. This did not go well. The bleeding did not stop for three days, so I had to go back to the hospital. And then?

"Your prostate is shaped like a Mickey Mouse cap," the surgeon said cheerfully, "and the cancer has spread to the edge of the 'ears.' We don't know if it has leaped beyond that or not."

There was a chance the cancer might be gone, lifted out with the prostate gland. But we could not know if it spread unless we waited — and we could not wait — because it might spread. Once in the bloodstream, they said, it could go anywhere.

So, radiation: the "GRRRRR-ZEEEET" of a huge machine zapping the area, every afternoon for an hour, five days a week, for six weeks.

And of course, "the shot." Lupron is a female hormone, which if put in a male may depress cancer growth. The first injection felt like being harpooned. The second and third shots were easier. This nurse knew better how the body worked, I think, because the medication slipped almost politely into the hip.

I had a quick lesson in the cost of health care. The bill (of which I paid nothing) was $960,000. Without insurance, the cancer could have done what it wanted, because I could not afford that much money — almost a million dollars?

"Now serving 245-a," said the robotic voice.

I went into the next room, rolled up my sleeve, thrust out my left arm. Supposedly it hurts less in the left, perhaps being closer to the heart.

As I always did, I told the nurse a story about my diving days at Marine World, when I held a dolphin's tail still for the vet to take a blood sample.

I asked him: "Since the veins on a dolphin's tail cannot be seen, how do you know where to take the blood?"

The answer is you take your thumb and root around on the gray tail flukes, till you find a loose spot on the skin, and that's where the vein is.

Then I closed my eyes, having done all I could to get the person with the needle on my side.

"All done," said the nurse.

"When will I know the results?"

Meaning when will I know if the cancer has come back...

Side effects? Urination has become an accuracy challenge. The male organ was literally shortened by the surgery, so there is less to grab and aim with.

There are also the joys of incontinence. I visit the senior section of the drug store frequently now, to purchase those padded adult garments nobody talks about.

Incontinence joke: "Do you wear boxers or briefs?" Answer: "Depends."

Overall, not fun. Still, as my wife says, "Consider the alternative ..."

Today, right now, that was what I am doing.

Prostate cancer is generally slow-growing, so even if the results came back bad, still I would have a couple of years.

But I need 20 years, at least; I am too busy to be dead!

I have to finish this book, of course, plus at least three others I intend to write, not to mention some plays and screenplays.

Roman and I want to walk our Katherine down the aisle, with all 3 of us walking.

Above all, we must pass Proposition 71, Part Two, so the California stem cell program can continue.

And there is one more worry, which only a parent can understand …

Three times a week I visit Roman's house, early in the morning, to help Roman with his shower and bathroom procedure. It is just something that has to be done.

But who will take care of him, when I am gone?

Question: did you read my previous book, *Stem Cell Battles: Proposition 71 and Beyond: How Ordinary People Can Fight Back Against the Crushing Burden of Chronic Disease*? It's a huge book, 432 pages, with different stories and lots of them.

If you did wade through it, you remember the great Stanford scientist Irv Weissman, and his work on the cancer stem cells.

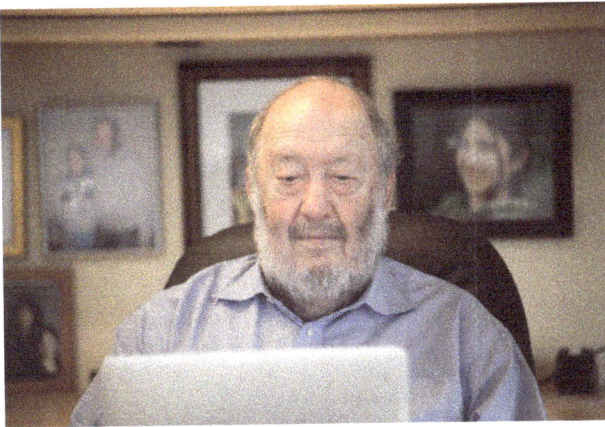

Irv Weissman (Christopher Vaughan).

The reason cancer is so hard to beat is that it has its own stem cells, the evil twin of the good ones. They are not detectable by the immune system, because there is a protein on their surface which acts like an invisibility cloak, or as Weissman puts it, a sign which says "Don't eat me."

You have heard of cancer survivors going into "remission" for a while, but then the cancer comes back? That means most of the cancer cells were killed, but not all.

If I am killed by cancer, the murderers are the cancer stem cells.

So Professor Weissman is trying to remove that cloak of invisibility, to erase the sign that says "don't eat me!" If he succeeds, the body's immune system should be able to find the cancer cells and kill them all.

That, I believe, is half the answer.

The other half is the part of the immune system that will attack the cancer.

It has a lovely name: Natural Killer (NK) cells. Often the secrets of science are buried under gigantic words. But Natural Killers? That I can grasp.

This morning, at nine o'clock, I would be talking with an expert on NKs: Dan Kaufman, M.D., Ph.D.

Dan Kaufman has used Natural Killers to wipe out cancer in lab rats. That may not sound like much at first, but I had seen before and after pictures of the rats. Some were so eaten up with cancerous tumors it was hard to look at them.

But there was also a different set of pictures: clean and healthy rats, beautiful. It was hard to believe they were the same animals.

Even when they were put to sleep and dissected, their bodies were completely clean, inside and out, the cancer wiped away ...

When I saw those pictures, I felt a click and shock, like recognizing a superstar. Could this be it, the answer to cancer?

I looked up Dr. Kaufman's contact information at the University of Minnesota, and said hello. Naturally the first thing I told him was that he should relocate to California, to a stem cell program with money for groundbreaking work like his.

He said no, at first.

But once a year or so I would call him up and remind him why California is the place he ought to be. I have no idea if I influenced his decision

Dan Kaufman (eurekalert.org).

or not, but now here he was at UCSD, one of the greatest stem cell research centers in the world.

"Hello, Dan," I said.

"Hello, Don," he said.

And then we were off on a talk where I constantly had to ask him to talk slower.

I knew he had tried for a CIRM grant, because I had fought for it. I read the agenda of the ICOC meeting where it would be decided. There was a list of grants being considered, no names. This was deliberate: if a grant was rejected, it was better not to embarrass the scientist publicly.

But I had recognized Dan Kaufman's research immediately. NK cells killing cancer? Not that hard to figure that one out, especially with him in California now.

I went to the meeting to argue for the grant, as any Californian can do. The grant required a score of 85, but the score beside Dan's grant was 75, below the cutoff.

But the public are allowed to comment. I disagreed with the judgement, and so when Chairman Thomas asked if there was public comment, I jumped to the mic.

The ICOC (Independent Citizens Oversight Committee, the board of directors) is 29 members: patient advocates, scientists, biomed folks. They make the decision.

I talked about the rats, so clean of cancer. I wished I had the before and after pictures to show them: nobody could look on them and not be moved.

I mentioned that the particular kind of cancer being attacked by this project was blood cancer, leukemia, the disease that killed my sister, and which still kills 20,000 Americans every year, and for which there was at present no cure.

The vote was taken. Of the 17 grants under consideration, 4 were funded: not his.

So my conversation today was just to be sure Dan would try again.

"Oh, yeah, absolutely," he said, "We're going to try for two. One for leukemia, the other for solid tumor cancers ..."

Solid tumor cancers, I thought to myself when he hung up.

Like the prostate cancer I might still have.

12 A POLITICAL OBSTACLE TO HEART DISEASE CURE?

Imagine something ridiculous.

You are standing on a high dock overlooking the ocean, and somebody falls in, and he can't swim. He is drowning before your eyes.

Fortunately, there is a coil of rope on the dock beside you, and you bend down to grab some and throw it down to rescue him when a big boot stomps on the rope.

"Sorry," says the owner of the footware, "It is illegal and immoral to save people with that particular kind of rope. You will have to go find a different rope."

Ridiculous? Yes.

But does it relate to stem cell research? Perhaps ...

A fight we thought was over may start up again.

As you may recall, President George W. Bush's stem cell policy was restrictive, allowing federal funding only for embryonic stem cell lines made before September 1, 2001. This was like telling the aeronautics industry they could only use airplanes in existence when the Wright Brothers flew at Kitty Hawk.

Fortunately, President Barack Obama reversed that short-sighted policy. My wife Gloria, son Roman and I were in that happy audience of scientists and advocates on signing day. It is now possible for the Federal government to support research on new embryonic stem cell research lines, not just the old outdated ones.

But on January 17, a new President took office, and nobody knows for sure how Donald Trump feels about stem cell research.

Will he see biomedicine as a benefit to every American?

Or will he side with the anti-research folks?

This matters. The National Institutes of Health (NIH) is a $32 billion government agency. If a state provides first funding of a new therapy, the NIH may contribute "add-ons," new and more substantial funding. For instance, the Roman Reed Spinal Cord Injury Research Act of 1999 provided roughly $17 million in state funding over ten years; but the NIH added to that — another $85 million!

If a new President discouraged NIH funding of embryonic stem cell (ESC) research, it could be devastating to all our hopes for research leading to cure.

On the state level, could California's stem cell program be at risk?

My gut instinct is that President Trump will understand, and will hopefully not attack ESC research. There has been so much progress; it would be wrong to reverse course, and deny the hope of cure to suffering children, women and men.

Freedom of science must be protected. That is why Proposition 71 and the California stem cell program was created, so research could be free to save lives and ease suffering, and not be blocked by political opinions.

Let there be no misunderstanding: California is no Wild West show; there are zero mad scientists running around doing crazy Frankenstein stuff.

Our researchers operate under stringent regulations: state laws, institutional review boards, as well as federal restrictions.

For example, where do our embryonic stem cells come from? They are made from microscopic fertilized and frozen eggs already scheduled to be thrown away: left over from the extras in the In Vitro Fertilization (IVF) procedure.

Here's how it works. When a childless couple seeks IVF assistance, about 15 sperm-egg combinations are made in a dish of salt water. The strongest one or two of these are implanted in the woman's womb; the rest are frozen for further reproductive use, given to another couple, thrown away, or donated to research. They are only eligible for research when the decision has been made to discard.

Important: Unless implanted in a mother's womb, these tiny dots of tissue, smaller than the period at the end of this sentence, can never

Deepak Srivastava (cirm.ca.gov).

become a human life. It is biologically impossible. No womb, no baby. This is not rocket science.

But what if those who oppose stem cell research for ideological reasons were successful? What could we lose, if embryonic stem cell research was outlawed?

Example: an embryonic stem cell procedure might rebuild a damaged heart from within.

That effort combines two of the world's greatest medical institutions, Gladstone Institute and Stanford University, and internationally-renowned researchers, Deepak Srivastava and Joseph Wu.

Their therapy is designed for patients with end stage heart failure: people imminently at risk of dying.

But let the experts talk; here are a few sentences from their grant proposal, answering questions from me.[1]

Question: how wide-spread is heart failure, and what are your odds of survival?

Answer: "The American Heart Association has estimated that 5.7 million Americans currently suffer from heart failure, and that another

---

[1] https://www.cirm.ca.gov/our-progress/awards/human-embryonic-stem-cell-derived-cardiomyocytes-patients-end-stage-heart-0

Joseph Wu (profiles.Stanford.edu).

670,000 patients develop this disease annually... Patients with end-stage heart failure have a 2-year survival rate of only 50% with conventional medical therapy ... worse than for patients with AIDS, liver cirrhosis, stroke and other ... diseases."

Question: could you live if your own heart was replaced with another?

Answer: "... (Heart) transplantation is highly effective at increasing patient survival, but is severely limited... by the very low number of hearts available."

Question: is there a way you could keep your own heart, and just have it repaired?

Answer: "... Our project seeks to treat patients by replacing their lost (heart muscle) cells with healthy ones derived from federally-approved human embryonic stem cell line WA07 ... human embryonic stem cells can be (turned into) heart muscle cells ... and be used to improve cardiac function following a heart attack."

Thanks to the California stem cell program, this amazing possibility is now being tested on pigs and lab rats.

Assuming continued success (and FDA approval), this therapy would be tested by offering it to people who already need the help of a heart machine: the Left Ventricular Assist Device (LVAD), which helps the heart pump blood.

After the cells are injected into the patient, success will be judged by gradually turning down the volume on the LVAD. When "the pump speed is turned down to minimal levels … (heart) function is assessed by echocardiography and the 6-minute walk test. Accordingly, it is possible to (judge) the effects of therapies without putting the patient at serious risk."

Will it work? We cannot know until it completes the rigorous FDA testing process.

But one thing is certain: If ideological ignorance gets in the way, and effective therapies are blocked, people like my wife may die of heart failure too early.

13 YOUR FRIEND, THE LIVER!

Best known for the Christmas classic, *It's A Wonderful Life,* starring James Stewart, Frank Capra also directed a series of educational science movies, including *Our Mister Sun,* and (on blood) *Hemo The Magnificent.*[1]

We need more such fun-but-accurate movies today: maybe one titled "YOUR FRIEND, THE LIVER!"

Until something goes wrong, we seldom think about this football-shaped three-pound purple organ, resting right beside the stomach. Incredibly versatile, the liver performs an estimated 500 chores: filtering toxins, storing blood and vitamins, working with proteins to help the body build muscle.[2]

Numerous liver diseases put this subtle machinery at risk. Just one, cirrhosis (scars on the liver) kills roughly 38,000 Americans every year.[3]

Often thought to be solely a result of alcoholism, cirrhosis can be caused by a variety of reasons, including virus infection, autoimmune hepatitis, and obesity.[4]

When the liver shuts down, bodily poisons are not filtered, and the owner dies.

At present, there is only one cure for liver failure: transplantation.

---

[1] http://www.avclub.com/review/frank-capras-wonders-of-life-hemo-the-magnificent-11767
[2] http://www.hopkinsmedicine.org/healthlibrary/conditions/liver_biliary_and_pancreatic_disorders/liver_anatomy_and_functions_85,P00676/
[3] https://www.cdc.gov/nchs/fastats/liver-disease.htm
[4] http://www.medicalnewstoday.com/articles/172295.php

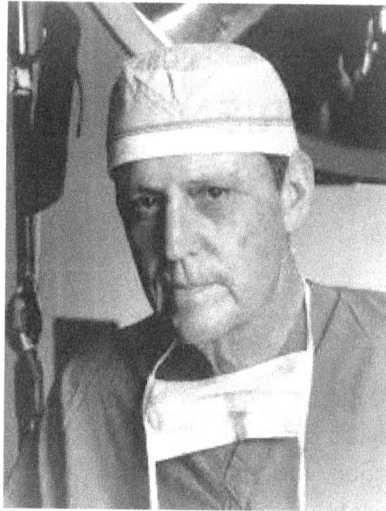

Thomas Starzl (en.Wikipedia).

First performed successfully by Dr. Thomas E. Starzl in 1967, transplanting a liver is a difficult, time-consuming operation. For years the only surgeon with this expertise, Dr. Starzl often had no back-up surgeon, and might be forced to continue operating as long as three days without sleep ...

The courage of that man! He loved dogs, but had to sacrifice 150 of them to come close to success. And, when he tried the procedure on people, the first five died. One bled to death on the operating table, the other four perished within three weeks. When similar attempts also failed in other nations, the rest of the world quit trying, because it seemed to be too difficult.

But Starzl did not quit. He also worked on transplanting kidneys with considerable success. He wrote two pivotal books on transplantation, one for each organ.

Nothing was easy. For next 12 years, half of Dr. Starzl's liver transplant patients died within a year after the operation.

Working with four European transplant centers, Dr. Starzl developed procedures and pharmaceuticals, until finally, in December 1981, U.S. Surgeon General C. Everett Koop called a conference to evaluate their work. When it was over, Starzl's procedure was declared a "clinical service," not an experiment.

Starzl's eminence may be judged by the staggering number of peer-reviewed publications (1,700) he authored, of which 461 concerned liver transplantation, still cited today. At one time, he was the most frequently-cited scientist in the world.

Today, largely due to his efforts, the liver transplant operation is considered survivable. The world owes Dr. Thomas E. Starzl a standing ovation.[5]

Even so, there are only about 6,000 liver transplants a year. Perhaps 30,000 people die waiting for a liver "match" to transplant.

My cousin Will and his wife Mary have been traveling back and forth across the country, from Oregon to Louisiana, for two years, trying to find a match for her liver. Her condition is PLD (Polycystic Liver Disease) where many cysts grow on the liver (brought on by a virus). So far, she is able to live with it, because there is still enough liver left to provide for the body's needs. But every time she gets sick, they worry, is this the end?

When successful, the results of a liver transplant can be wonderful.

Example: one of the many different kinds of liver diseases is the oddly-titled Maple Syrup Urine Disease, so called because if an infant has the condition, their diapers smell like pancake syrup. The condition is a subtle torture. In the Greek legend of Hell, Tantalus was condemned to watch others feasting while he starved eternally, food always just out of his reach, giving us the word "tantalize."

Now, imagine a 6-year-old girl, Vanessa Lupian, unable to eat normal food, including protein: no meat, milk or eggs. And it was worse than just an enforced Vegan diet. She not only had to eat a very specific diet, but not much of it. Every meal was weighed on scales. If she ate "wrong," she faced hallucinations, and incessant vomiting, while her parents feared possible brain damage and death.

But a matching liver was found for Vanessa Lupian: (actually two, because the first one did not "take") and after four operations she

---

[5]http://www.starzl.pitt.edu/transplantation/organs/liver.html

was able to eat normally for the first time in her life. Her favorite food? French fries.[6]

Even so, there may be a better way.

Problem: A natural glue called collagen keeps the liver together. That's fine. But too much collagen becomes tangles of scar-like fibrosis, choking off the liver's healthy cells, leading to slow death.

CIRM-funded UCSF Professor Wilger Hollenbring is approaching liver disease from a "Midas touch" perspective. The mythical king could turn lead into gold — what if the deadly scars of liver disease could be transformed into useful cells?

Hollenbring uses a form of stem cell therapy called "direct reprogramming."

Using natural chemicals like those which change embryonic stem cells into hepatocytes (liver cells), he is attempting to turn one kind of cell into another.

Is it possible for the fibrosis "scars" to be turned into healthy liver cells?

Dr. Willenbring, and partners from Germany and China, are trying to do just that.

They hope to get rid of the scars, which block the function of the liver, and replace them with normal healthy liver cells: hepatocytes.[7]

May your liver never falter in its natural filtering job![8]

[6]https://blog.cirm.ca.gov/2012/09/17/liver-transplant-saves-girls-life-but/
[7]https://blog.cirm.ca.gov/2016/06/02/good-from-bad-ucsf-scientists-turn-scar-forming-cells-into-healthy-liver-cells/
[8]http://www.cell.com/cell-stem-cell/abstract/S1934-5909(16)30089-3

14 "BRING 'EM BACK ALIVE"

World-famous hunter for zoos, and author of the book, *Bring 'em Back Alive*, Frank Buck once had a request for a truly huge king cobra.
The snake was caught, placed in an old wooden box, and brought to Buck on his ship. However, in the jungle's humid weather, the boards were decayed and weak.

A better travel crate was built, and the collector was helping lift the old box to transfer the snake when the rotten wood gave way.

The cobra burst free, thumping onto the deck. Furious, it reared up, hissing, expanding its hood. All but one of the humans fled for their lives, climbing up on stacked boxes.

But Frank Buck was not about to lose his snake. He yanked off his jacket and hurled himself forward, pinning down the snake. Or so he thought.

Because the fight was not over. The cobra struggled violently to work out from under the jacket. To Buck's horror, he felt a thump! on the back of his leg. The cobra's venom is so virulent — and so much of the poison is injected through the chewing bite — a cobra can literally kill an elephant. Instantly the collector felt the first symptoms of death by cobra venom.

As he hung on for dear life, the helpers climbed down from their various perches, and helped Buck recapture the reptile. Only then did Frank realize he had not been bitten, but only struck by the writhing cobra's tail.

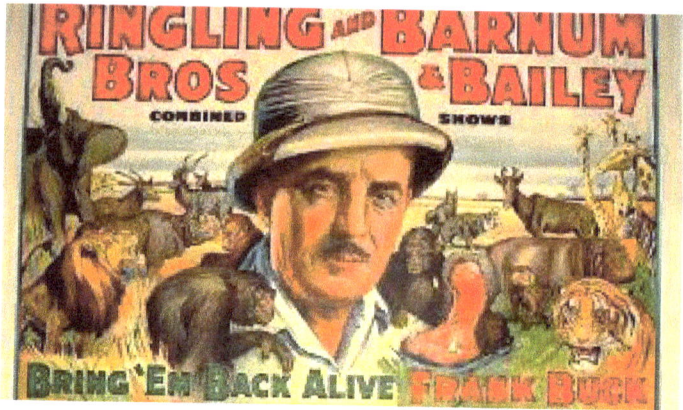

Frank Buck (http://www.mnn.com/earth-matters/animals/stories/).

Delivered to Raymond L. Ditmars, curator of the Chicago zoo, the snake became famous as the world's largest cobra: 13 feet, six inches long, two times the height of a man ....[1]

Now: imagine a death even more terrifying: Huntington's disease (HD). A cobra's venom may kill in moments. Huntington's may drag out 20 years in paralysis and slow death, but first the disease has a mental effect.

"We had to keep telling ourselves, this is not Dad, it is the condition ...," said family member Sherry (last name withheld). Her father had once been kind and gentle, but now became depressed, short-tempered and occasionally violent.

"You know how it is when you are a teenager. If somebody says, 'get the dishes done,' you might not jump and do it instantly. He would grab us and (hit) us."

Gradually, his abilities were stolen. He could no longer hold a job. Driving a car was out of the question. He could not even speak properly, after his tongue lost its coordination. Standing and walking were denied him as well. If the kids wanted to take a picture with their dad, they had to climb onto his home hospital bed.[2]

---

[1] http://newfpl.weebly.com/uploads/5/1/1/0/5110912/king_cobra.pdf
[2] https://www.cirm.ca.gov/our-progress/video/spotlight-huntingtons-disease-2010-sherry

Leslie Thompson (faculty.uci.edu).

And the worst of it? A 50% chance a Huntington's may be passed to the children.

There is a test for a young person to diagnose for Huntington's, but some choose not to take it: not wanting to know if they have the disease.

Part of this is for a very practical reason. If insurance companies know you have the disease, they may deny you coverage. This is why it was such a big deal when Affordable Care Act (ACA, or Obamacare) said **no one could be denied coverage for a pre-existing condition**. This great requirement not only saves a family from financial ruin, but also gives researchers a better chance to find a good therapy. Secrecy is the enemy of cure; if diagnosis is delayed, how can we know what it looks like in its early stages, when it is perhaps more easily treated?

Remember the snake poison, which Frank Buck was mercifully spared?

Cobra venom is a protein: not the kind which builds muscle, but one which brings death. But there is an antidote. Taken in time, it can save a life.[3]

---

[3] http://news.nationalgeographic.com/2016/05/160503-snake-bite-antivenom-asia-africa-animals/

Huntington's disease is triggered by the oddly-spelled huntingtin's, protein, which is "... under investigation as part of Huntington's disease clinical research ..."[4]

Could there be an antidote for the cause of Huntington's?

I took my question to Dr. Leslie Thompson, a world-renowned authority on Huntington's disease. (Note: for more on her work, see Chapter 44 in SCB)

Head-quartered at UC Irvine, CIRM-funded Dr. Thompson is using both direct programming and embryonic stem cells to "provide neuroprotection to the brain (as well as) the possibility of cell replacement."[5]

The direct programming? Using skin cells taken from people with Huntington's, and a cocktail of chemicals, Dr. Thompson and her colleagues made an HD stem cell line. With this "disease in a dish," it was possible to try different medicines on the cells, see what works, without endangering a human.

The second part involved embryonic stem cells, which would be changed (differentiated) into nerve stem cells. If these healthy cells were transplanted, they might protect patient nerves from being destroyed by Huntington's, perhaps blocking the protein that causes HD harm.[6]

To prevent further damage from being done, or even delay the start of the disease? That sounds pretty much like an antidote to me.

If it works, this approach may help defeat HD and other nerve-damage diseases like ALS (for which her colleague Clive Svendsen just received FDA permission to begin human trials), Alzheimer's, or Parkinson's disease.

A 27-year veteran of the fight against HD, Dr. Thompson has gathered a team of experts: Mathew Blurton-Jones, Ed Monuki, Steven Cramer, and Neal Hermanowicz of UCI; Michael Levine and Marie-Francois Chesselet of UCLA; Clive Svendsen of Cedars-Sinai Medical Center; and Gerhard Bauer of UC Davis.

---

[4] https://en.wikipedia.org/wiki/Huntingtin
[5] https://news.uci.edu/health/uci-team-gets-5-million-to-create-stem-cell-treatment-for-huntingtons-disease/
[6] http://curehd.blogspot.com/2012/01/advocacy-pays-off-huntingtons-disease.html

Gerhard Bauer (blog.cirm.ca.gov).

Ed Monuki (Faculty.uci.edu).

"The group is partnering with Terumo BCT, a global biotech company that employs a system for expediting the growth of stem cells. BioTime, Inc., (a company begun by stem cell pioneer, Mike West) … is providing the … cells." These will be developed into neural cells at the Good Manufacturing Practices (GMP) facility at UC Davis.[7]

---

[7] https://news.uci.edu/health/uci-team-gets-5-million-to-create-stem-cell-treatment-for-huntingtons-disease/

Matthew Blurton-Jones (faculty.uci.edu).

The work will be headquartered at the magnificent Sue and Bill Gross Stem Cell Research Center at UC Irvine.[8]

Dr. Thompson warned me not to get too excited, and I understand her scientific caution. All it takes is one small detail to be wrong, and an entire therapy may not work. But we have top-notch people, a powerful approach to stem cell research, application to other diseases, and, at this moment in time, the funding is there.

No one can predict the path of science. But I know what my heart tells me.

If the California stem cell program remains funded and free, I believe we will soon have a very different prognosis for folks with Huntington's.

As Frank Buck might say, we will "bring 'em back alive" — and well.

---

[8] http://stemcell.uci.edu/

15 THE COLOR OF FAT

Why is it some people do not gain weight, even if they own stock in an ice cream parlor? These irritating people never seem to exercise or diet, and yet they stay lean and healthy. What gives them such an unfair advantage?

It may be the color of their fat.

Every adult human has a mixture of white, brown, and beige-colored fat.

Brown fat burns off quickly, as heat and energy; the athlete's chiseled body means he/she has more brown fat. Hibernating bears and newborn humans also have more brown fat, for warmth. Infants cannot shiver, so their body thermostats are set up for heat. The beige fat is found in narrow threads; more on that later.

And we who fight the battle of the bulge? White fat is blubber, which mostly just stays stored around the middle: the gut or butt. When we "pinch an inch" on the waistline, that's white fat. And when we clutch a handful, that's obesity.

Does it matter, if we are thin or plump?

Unfortunately, it is life and death. As our weight goes up, so do our chances of Type 2 diabetes, heart disease, kidney problems, colon cancer, and stroke.

Gloria and I went to a food class last week. Every person in the room was at least mildly overweight, except the instructor. At 226 pounds, I was no exception.

During the class, the instructor looked at me and said, "You look so confused ..."

I was not confused, just overwhelmed. What she described seemed impossible. She had a pile of plastic models of various foods, and would throw them on a plate to show us how to eat. Half of the plate was for vegetables. The other half was divided into three parts, tiny little dabs of what I considered edible food.

Her idea of a big treat was a teaspoon of raisins.

Afterwards I went online and bought several cheap books on diabetes diets, and some of them didn't seem quite so impossible. My wife is inclined to be vegetarian anyway ... could we bring our weight down, adjust Gloria's diabetes blood count, make a permanent adjustment to our life style?

The odds are against it.

To live endlessly self-starving lives, hungry all the time, counting the calories in every bite? It is possible, as proven by those who fight the deadly Type 1 diabetes, where the choice is be careful or die; but most of us with weight problems just stay that way. If we lose a couple pounds, we put it right back on.

Look around you at the gym. These are people who have made a conscious commitment to lead healthy lives, and even so most of us are overweight. I go to the gym three times a week, and I still weigh too much.

Shingo Kajimura (ucsf.edu).

It is not our fault, my mind whimpers, I'm fat because the world has changed!

When I was a skinny high school freshman in 1959, we had one car, which my Dad took to work. I walked three miles to school, and three miles back, forced to exercise, and there were no fast foods. I distinctly remember my first hamburger. It was called a "butterburger," delicious and deadly. When good-tasting food like that became widely available, and more cars to cheat us from exercise, we ate too much and did too little, and a nation became fat.

There was one overweight girl at my high school, who must have weighed about 300; I had never seen anyone that big, so young. I was not rude to her, of course, but why was she built so differently? When she bounced on the trampoline in gym, it sagged almost down to the floor.

Today, people that size are not uncommon. One Californian in four (26%) is obese: dangerously overweight. And California is more health conscious than some states; nationally, the obesity index is about 35% — one in three.[1]

Financially? The cost of treatment for obesity-related ailments is nearly $190 billion a year, roughly 20% of all medical costs ...[2]

But what if we could change white fat to brown?

There are scientists trying to do just that.

One is Shingo Kajimura. He works at the Eli and Edythe Broad Center of Regenerative Medicine, at the University of California at San Francisco.

Dr. Kajimura's lab has a stated goal: "Engineering fat cells to fight obesity."[3]

"Brown Adipose (fat) Tissue (BAT) acts ... as a natural defense system ... against the development of obesity... (which) has prompted

---

[1] https://www.niddk.nih.gov/health-information/health-statistics/Pages/overweight-obesity-statistics.aspx
[2] https://www.hsph.harvard.edu/obesity-prevention-source/obesity-consequences/economic/
[3] http://kajimuralab.ucsf.edu

our lab to examine the links between BAT, obesity, and metabolic disorders."[4]

Remember those beige cells, threaded among the blubbery white fat?

"For the first time, a research team, led by UCSF biologist Shingo Kajimura, has isolated energy-burning "beige" fat from adult humans, which (may enable us) to convert unhealthy white fat into healthy brown fat..."[5]

The beige fat can apparently "recruit" (change) white fat cells to brown, or least give the white fat the ability to more easily burn off, as if it were brown.

We could lose weight more easily, live longer, healthier lives ...

Shingo had (like me) an ocean background, growing up fishing almost every day, to supplement the family's food supply. I called him for a phone interview, and it was hard not to spend our time talking about fish and sharks.

"I wondered how animals adapt to a severe environment," he said, "Humans use shivering as a way to generate heat, but animals don't."

"In early 2000, only four or five labs were studying brown fat. One was led by Dr. Bruce Spiegelman, I was a research fellow under his leadership for several years."

"In 2009, two major discoveries changed fat research: first, it was found that humans kept small amounts of brown fat from infancy into adult years, and that was correlated with (how fat they were). If they were lean, more brown fat. Second, beige cells (could) recruit white fat into brown, or at least act like brown."

"We found a way to turn skin cells, muscle cells and white fat cells into brown fat."

"Since most people are reluctant to undergo transplant surgeries, we are currently trying to (develop) an injection or a pill, to help a patient fight obesity by turning white fat into brown." — Shingo Kajimura, personal communication.

---

[4]http://www.cell.com/cell-metabolism/references/S1550-4131(16)30378-3
[5]http://www.theglobeandmail.com/life/health-and-fitness/health/california-researcher-wants-to-fight-obesity-with-fat-cells/article28220440/

Another scientist with a long history of anti-obesity research is Chad Cowan from the Harvard Stem Cell Center, run by Brock Reeve, that superb organizer and fund-raiser, brother to the late stem cell champion, Christopher "Superman" Reeve. In a 2014 experiment, Cowan used stem cells as a testing device: trying to find out if there were drugs available which might change white fat into brown, or at least make the white fat function as if it was brown.

With the help of a pharmaceutical company, he ran through roughly 1,300 drugs. He found two that made the desired change, but they had dangerous side effects. Even so, the idea is valid; a larger pharma company would have access to millions of drugs, and surely some would be useful.[6]

Kajimura speaks highly of Cowan, crediting him with turning embryonic stem cells into brown fat, and repeating the effort with induced pluripotent stem cells.

If I had my way, scientists like Cowan and Kajimura would be locked in a laboratory, having a contest. They would not be allowed to come out (sorry, guys!) until one of them found a cure to obesity, with a huge money prize as incentive.

Well, maybe my idea is not completely practical! But, there is precedent for a massive money prize, called a Social Investment Bond (SIB). The idea is if somebody makes a positive change for society, they get a percentage of the savings it brings to the country, for an established length of time.[7]

So if Shingo or Chad found the answer to obesity, they would automatically become rich, and never have to worry about grant money any more. We do not have SIB grants in America yet, but the idea seems worth considering.

But even in our current system, if a person came up with a way to make everybody leaner and more fit, without nearly-impossible life style changes in diet and exercise, they should do well financially …

I wish them (and all cure research scientists) the very best of luck.

If they win, we win.

---

[6] http://news.harvard.edu/gazette/story/2014/12/a-pill-to-shed-fat/
[7] https://en.wikipedia.org/wiki/Social_impact_bond

16 REVENGE FOR MY SISTER

It is more than half a century since leukemia stole my sister — at the age of 23.

If she had lived … Even in the late fifties, Patty understood the promise of computers. She carried around stacks of cardboard rectangles with holes punched in them, and loved to talk about binary codes. She was the gentlest soul I ever knew, sharing kindness like sunlight. Not a day goes by that I don't miss her.

But if she was alive today, and had just received that terrible diagnosis, things would be different. Today, we know how to fight.

I would get her involved in a clinical trial. How?

First, at the website, **www.clinicaltrials.gov,** there is a search box at the top of the page. Typing in the word "leukemia" tells where experiments are being done: NOTE: these are *not* all FDA-approved clinical trials. After choosing one, I would go to the **International Society of Stem Cell Researchers (www.ISSCR.org)** and see if there was anything good or bad about the scientists involved. I would search the website, and ask for the ISSCR booklet on how to choose a stem cell therapy.

Here is the title of one promising study: "UC-961 (Cirmtuzumab) in Relapsed or Refractory Chronic Lymphocytic Leukemia." Let's say I chose that one.[1]

Using the information on the page, I would contact the leader of the trials, in this case Catriona Jamieson, MD, PhD, of the University of California at San Diego, and ask if my sister fit the parameters for

[1]https://clinicaltrials.gov/ct2/show/NCT02222688

Tom Kipps (ucsdnews,ucsd.edu).

John Dick (https://www.uhnresearch.ca/researcher/john-e-dick).

inclusion in the clinical trials of Cirmtuzumab, a stem cell-derived drug to fight leukemia. If Patty had been one of the participants, she would have driven to San Diego, and taken part in the battle.

What is this particular project?

Leukemia has cells with a marker called ROR-1, which makes the disease grow fast, "metastasizing," which is what finally kills the patient.

The hope is that a new drug, Cirmtuzumab, will destroy ROR-I, and slow or stop the leukemia. Look at that name: Cirm-tu-zu-mab, named

Randy Mills (blog.cirm.ca.gov).

after CIRM, the California Institute for Regenerative Medicine. I find that bold and beautiful.

Dr. Jamieson and her co-workers Dennis Carson, Tom Kipps, and Canada's John Dick have been working together for many years to defeat leukemia.[2]

But there is more: a new approach at fighting. To understand its significance, consider how the ancient Romans conquered the world.

The Romans fought with unity and structure, exemplified by a military tactic called the shield wall, or tetsudo (turtle). To see this in action, check out the artwork of Wenceslas Hollar of the University of Toronto. The soldiers' shields overlapped, both on the sides and overhead, so arrows bounced off like rain off a roof.[3]

Usually outnumbered, the Romans fought as a unit, and they won.

Similarly, CIRM fights in an organized way. One example of their teamwork approach is the Alpha Stem Cells Clinical Network.

[2] http://www.bloodjournal.org/content/126/23/1736?sso-checked=true
[3] https://en.wikipedia.org/wiki/Testudo_formation

"CIRM has funded three stem cell-focused clinics, housed within existing medical centers. This network of clinics will attract and conduct — high quality trials ..."[4]

There are three such Alpha Clinics right now, soon to be four, each with a different disease focus. At the City of Hope hospital they are taking on brain tumors and HIV/AIDS; at the consortium of UC Irvine and UCLA, the challenge is blindness, sickle cell, cancer and "Bubble Baby" disease; UCSD is going after Lou Gehrig's disease (ALS), diabetes, and leukemia.[5]

To help the process along, there are two backup partners — the Translating and Accelerating Centers — whose job is to un-complicate the process for the scientists.

CIRM President Randy Mills put it well, saying:

"Many scientists are brilliant researchers but have little experience running a clinical trial; (now) they don't have to develop those skills (because) we provide them ... they are free to do what they do best, namely science."[6]

The Translating Center works with the scientists to get them ready for the Food and Drug Administration (FDA), including the very first step, the Investigational New Drug (IND) application. No IND, no clinical trials and no new therapies.

The Accelerating Center helps develop tests to prove the safety and efficacy of the method being tested. A highly respected company, Quintiles, Inc., was hired for this purpose.

If the clinical trials could be made more systematic, what a useful time-saver that would be. "Roman's Law" had formerly paralyzed rats walking In 2002, and 15 years later, we are still going through test after test after test!

But with Alpha clinics, we would have test protocols already in place, ready to go, to "accelerate therapies to patients with unmet medical needs," as CIRM says.

---

[4]https://www.cirm.ca.gov/patients/alpha-clinics-network/about
[5]https://www.cirm.ca.gov/patients/alpha-clinics-network/alpha-clinics-trials
[6]https://blog.cirm.gov/2016/06/15/accelerating-the-drive-for-new-stem-cell-treatments/

The Alpha clinics will also be a place of trust: what Dr. Jamieson calls "a one-stop shop for patients," both for accurate information, and perhaps to sign up to be part of a reliable and regulated clinical trial.

It will be an answer to the phony and dangerous "stem cell tourism" rip-offs, where people have their money stolen in exchange for fake or untried "stem cell" treatments, which may be just worthless saltwater injections, or worse. Too many people are paying $40,000 or more for something that may be called a clinical trial, or even labeled as an outright cure, but which could endanger the patient instead of healing them, as well as tarnishing the image of honest regenerative medicine.

All this is too late to help Patty Reed, of course. But the Alpha Clinic Network approach may help save other lives, equally precious.

I hate leukemia: that miserable disease which killed my sister. I wish I could clutch it, choke it, and physically destroy it.

But in a way, we can. For is not cure the only real killer of disease?

ALSO: The ISSCR website has a great handbook for considering stem cell treatments: **http://www.isscr.org/home/publications/patient-handbook**

17 A STORY WITH NO HAPPY ENDING?

I do not like the movie "ROGUE ONE: A Star Wars Story" because of its ending. Spoiler alert; everybody dies. The hero and heroine get exactly one hug, holding each other in their war clothes on the beach, and then the line of heat approaches, as the Death Star explodes the planet.

My 12-year-old grandson Jason explained to me how the ending was necessary because of the story line of the first movie, etc., and "Rogue One" was hugely popular, so it worked for others. But for me it was too sad. I do not go to movies to leave with a crushing sense of failure, like, well, we just lost the entire world!

There is enough pain in real life.

Like the email to Gloria which read: "You have diabetes."

I wondered if I should tell Bob Klein, the man who began Proposition 71, the citizen's initiative which led to the California stem cell program.

Maybe not. Less than one year ago, his beloved son Jordan passed away from complications of Type 1 diabetes. He was only 26 …

I was in my little office at Americans for Cures Foundation, where I work three days a week. This is a place of peace for me. I sit with my computer and write about stem cell stuff. The door is closed mostly; I am a writer, alone is what we do.

But anyone that wants to can slide the door back, as just now it did.

It was Bob.

I told him the latest stem cell gossip (Dan Kaufman of cancer-killer fame just got a grant from CIRM, as did Leslie Thompson to fight Huntington's!) and Bob suggested I reach out to a friend in the Huntington's community and thank her for all her hard work. (She likes to keep her name private, so I will.)

ViaCyte and JDRF (Viacyte.com).

He slid the door shut. I wasn't going to tell him — I wasn't —
"Bob?"

The door slid back.

"Gloria has diabetes."

He winced. But before I had time to regret telling him, he said:

"Has to be type 2, at your and Gloria's age. It's manageable, diet and exercise. Vegetables, lots of vegetables, remember?"

Just the day before, he had gently nagged me about my low regard for edible plants. Working at Americans for Cures means a free lunch every day, and Bob had asked the lady who orders the food to find out what kind of vegetables I liked. When she questioned me, I could not think of any, except corn, which I am told does not count. Straining, I said I liked vegetables cooked in a wok, and also shepherd's pie.

Bob smiled, told me to keep him posted, and walked back down the hall to the waiting piles of paperwork, and a meeting schedule the Pope would object to.

Mimi Gardner, redheaded office worker and friend, poked her head in:

"The important thing is, it's not the end of the world. My husband has it."

Gloria's life depends on a leaf-sized organ, the pancreas. When healthy, the pancreas produces insulin to balance the blood sugar level.

Francisco Prieto (cirm.ca.gov).

If the pancreas does not work, and the blood sugar level gets too high or too low, terrible things can happen. One of Gloria's family members died of diabetes; others have had limbs amputated.

I wrote to Dr. Francisco Prieto, the representative for diabetes on the CIRM board of directors. He responded immediately, saying:

"Hi, Don: The important thing is that, thanks to science we have lots of tools to control diabetes now. I assume she has Type 2 (more than 90% of all cases) so that means she likely will not need to inject insulin... Some people with Type 2 do end up needing to use insulin or other injectables, but that depends on how severe or advanced the diabetes is ..." — Francisco Prieto, M.D., personal communication.

An email to her doctor revealed Gloria does indeed have Type 2 diabetes, the lesser of two evils. Type 2 is dangerous, no question. But if Gloria and I eat carefully, exercise often, and stay in touch with the doctor, life can go on; her blood test number (6.5, where diabetes begins) can hopefully be controlled.[1]

But if we do not take it seriously, Type 2 can lead to coma, blindness, kidney failure, limb amputation, and death.

---

[1]http://www.mayoclinic.org/diseases-conditions/type-2-diabetes/diagnosis-treatment/diagnosis/dxc-20169894

Mimi Gardner (personal).

Type 1 is worse: a life of pain and medical attention. Monitoring blood sugar levels means poking your finger to take blood — perhaps several times a day — or wearing a monitor for that purpose. Every mouthful of food must be evaluated for its sugar/starch content. Too much or too little blood sugar can be deadly.

Insulin must be injected in a calculated amount. Although this keeps people alive day to day, still a diabetic's lifespan may be ten years less than normal.

What can be done to cure diabetes?

A donated pancreas can be transplanted.[2]

But there are never enough to go around. There must be a suitable cadaver pancreas from a person who signed up to donate body parts after death.[3]

If you are interested in being an organ donor, as I am, a contact is below[4]:

CIRM is supporting numerous stem cell attempts to cure diabetes.[5]

---

[2]https://en.wikipedia.org/wiki/Pancreas_transplantation
[3]https://www.unos.org/transplantation/
[4]https://www.organdonor.gov/index.html
[5]https://www.cirm.ca.gov/our-progress/disease-information/diabetes-fact-sheet

The most exciting approach was developed by ViaCyte, Inc. CIRM has backed it substantially, over $50 million in grants and loans over the past decade. CIRM's assistance is helping ViaCyte navigate the "valley of death," the shortfall of funds in the lengthy series of clinical tests.

As you know, embryonic stem cells can become any cell in the body. They first become a precursor, (the in-between stage), and after that the cell desired.

What if a vial of stem cells could go from embryonic to precursor to beta cells and duplicate the action of the pancreas, providing insulin when it is needed?

That is half of the ViaCyte attack on diabetes.

The other half is a container with holes in it, the "Encaptra"® delivery device.

Flat, thin, and about half the size of a credit card, the combined device and contents (PEC-Encap™ product candidate) is placed under the skin. There, it is hoped, it will integrate with the body. The precursor cells inside will turn into beta and other cells needed; the body will provide nourishment — and the holes in the device are too small for the patient's immune system to enter and kill the cells.

The insulin "dispenser" can be taken out if needed, but the idea is that it will be in the body on a long-term basis, and its owner can just forget about it.

There is also a variation with a more open dispensing device. Intended for those in greatest danger, the PEC-Direct ™ product would have the same pancreatic cell ingredients; the primary difference is in the dispenser. In this case, the patient will require immunosuppression therapy to protect the implanted cells.[6]

Is this just pie-in-the-sky-maybe-someday stuff? No. The ViaCyte PEC-Encap ™ candidate is being tested right now, with PEC-Direct ™ almost certainly soon to follow, with patient volunteers who want a normal life again.

ViaCyte, CIRM, JDRF and others are challenging a gigantic disease.

Estimates of the medical costs of diabetes? As high as $245 billion a year — a quarter of a trillion dollars ...[7]

---

[6] http://viacyte.com/products/pec-direct/
[7] http://www.jdrf.org/austin/wp-content/uploads/sites/3/2013/09/JDRF-Fact-Sheets-2013.pdf

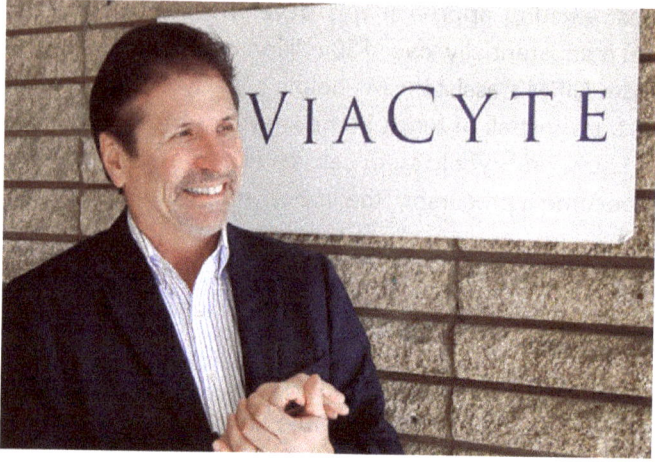

Paul Laikind, ViaCyte President and CEO (Viacyte.com).

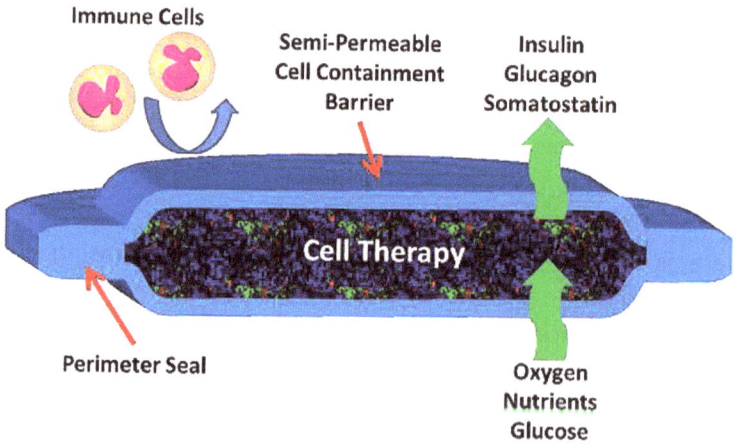

ViaCyte's anti-diabetes device (Viacyte.com).

But the size of the fight must not deter us; rather, it should underline the challenge, make us eager to take it on; directly or indirectly, diabetes threatens us all.

We are the writers of our lives; let us write a happy ending to the diabetes story.

18 AGING AND STEM CELLS

"The wonderful one-hoss shay … lasted one hundred years and a day …."
Poem by Supreme Court Chief Justice Oliver Wendell Holmes

My grand-daughter Katherine, age 7, once asked me why I am so much older than Grandma, since my hair is white and Grandma's is dark brown.

I told her I was middle-aged, just 71. She laughed and went about her business. But I was actually quite serious.

My father, Dr. Charles H. Reed, is still going strong at age 95. He plays tennis three times a week, and does a weightlifting routine devised for him by a Navy Seal. Mentally, he is active as well, currently reading through the Bible in French. (He speaks 12 languages.) He recently went to Morocco, where he shared his religious convictions with those of a very different faith. I worried he might have a missionary-style conclusion to his life, but he came home in one piece.

I want to age spectacularly. By that I do not mean to suddenly wrinkle, rot and explode all at once, like Supreme Court Chief Justice Holmes' wonderful one horse shay, (buggy) which blew up on its hundredth birthday, but just to stay in good shape, mentally and physically, until I die at last with my chores done.

I hope to skip some of the horror-show age-related diseases, like Alzheimer's, macular degeneration, cancer, heart disease, atherosclerosis, arthritis, and so on.

So how about a century of active healthy life, with faculties diminished but their function still intact — is there a way to achieve that?

Is there anything we can do, beside the obvious (exercise, diet, avoidance of stress) to maintain our youthful vigor? Maybe so.

Dave and Charles H. Reed (age 95).

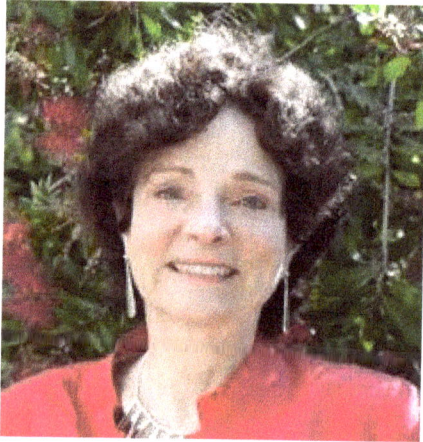

Helen Blau (Photo by Amparo Garrido).

Dr. Helen Blau of Stanford has numerous ideas (backed up by research) on how to make a healthy long life possible.

One way is to stay strong by maintaining muscle mass. As an old-time weightlifter that appeals to me, even now when I can only lift "baby weights." Just this morning, muscle mass saved me from falling down a steep flight of stairs. I tripped, lost balance, was on my way down toward a plate glass window, but crouched and caught myself; I still retain some leg strength. Balance alone is not enough; you need muscle to catch yourself, when you start to tip over.

Dr. Blau studies muscle stem cells (MuSCs) which she states "are responsible for the maintenance and regeneration of … muscle mass, crucial to mobility and quality of life. With advanced age, the proportion of (functional) muscle stem cells … declines." She hopes to "target and expand the …. MuSCs in the aged muscle tissues, enabling … muscle repair in the elderly."[1]

In addition to large skeletal muscles, Blau is concerned about small useful ones: like those around the eye, or those which grip the hand, or close the sphincter.

At the Baxter lab at Stanford, she is working on something spectacular — attempting to "physically lengthen …. the telomeres — the caps on the end of chromosomes that protect the cells from the effects of aging." She and Dr. John Ramunas are trying to turn back the clock on aging cells …

"Extending the life of cells by preventing — or reversing — the shortening of telomeres … (could) lead to new (ways) to treat age-related diseases.

"Now we have found a way to lengthen human telomeres … turning back the internal clock in these cells by the equivalent of many years of human life," explained Blau.

Remember your biology classes about chromosomes? They look like a medicine capsule, with little stringy telomeres sticking out each end. When the body is young, telomeres are long. With age, they shorten. It may be possible to lessen some aging effects by influencing telomeres with an enzyme called telomerase.

"Telomerase is normally present in high levels in stem cells, but drops off once the cells mature. Blau's modified RNA gives the aging cells a shot of telomerase, after which they begin behaving … half their age …"[2]

---

[1] http://web.stanford.edu/group/blau/
[2] https://blog.cirm.ca.gov/2015/01/28/extending-the-lease-stanford-scientists-turn-back-clock-on-aging-cells/

"We have … a way to lengthen human telomeres … turning back the internal clock in these cells by the equivalent of many years of human life," (said) Blau.[3]

To say we wish her success is an understatement. Look around.

America is aging. How many people are 65 and older? 46 million, 14% of the total U.S. population. If we age with strength, we retain more dignity, and won't need to depend on attendant care so much.[4]

Otherwise? Ask young people if they are looking forward to being care-givers to their parents …

We do not need to be a nation of feeble geezers. We can age with strength. Movie star Kirk Douglas is 100 years old right now, and he still exercises every day, and writes poetry and novels.

And if we live our days productively, we will never be bored, as exemplified by President/soldier Ulysses S. Grant.[5]

The impoverished ex-President, dying of cancer, wanted to be sure his wife Julia would be provided for. He also wanted to set the record straight on such matters as the Mexican-American War, which he fought in, but felt was a crime. He asked Mark Twain to help publish his book, and Twain said if the former President could find the strength to write it, he would make sure it was published.

Grant wrote the book, finishing it in bed. When it was done, he sighed, closed his eyes, and died.

Twain published the book, financing it with his own money. He paid veterans to act as salesmen for it, raising $450,000 for Grant's widow. Relinquishing all royalties, Twain did not take a penny of it, though it nearly broke him financially.

May all our lives be so full and long, and end with such nobility.

---

[3]http://med.stanford.edu/news/2015/01/telomere-extension-turns-back-aging-clock-in-cultured-cells.html/
[4]https://aoa.acl.gov/Aging_Statistics/Index.aspx
[5]https://en.wikipedia.org/wike/Personal_Memoirs_of_Ulysses_S._Grant

19 THE "IMPENDING ALZHEIMER'S HEALTHCARE DISASTER"

"I'm just going to the Ladies' room," said Gloria's friend C., "I'll be right back," she added to her husband M., who smiled and nodded.

All seemed normal in the crowded restaurant. Gloria kept talking to C.'s husband M., and he would cheerfully respond in what sounded like a foreign language.

Then he rose abruptly, scooted back his chair, and unbuckled his pants.

"No, no!" said Gloria, running around the table to pull M.'s pants together.

M. slapped her hands away, repeatedly, all the time talking rapidly and incomprehensibly. It got very quiet in the restaurant, as Gloria and M. fought for control of the zipper. And then —

"What are you doing to my husband?" snapped C.'s voice.

"No, no, I was just trying to — he was undressing and…" Gloria tried to explain.

C. burst out laughing.

"Oh, I know, I know!" she said, as the two women worked to calm M. down, rearranging his clothing.

Her husband had Alzheimer's disease.

She did not want him institutionalized, and kept him at home as long as possible, trying to provide for his needs herself. When he became too frightened, M. might clutch tight to her, so she had bruises on her arms. C. was loyal to a fault; but when M. wandered away one night and got lost, she finally accepted there was no choice but a managed care facility.

When he was in "the home," C. stayed by M.'s side all day, every day. She even hired someone to sit with him at night, worried the institution might tie him to the bed, to prevent him from trying to leave.

When he was hospitalized for end-of-life care, C. still was beside him.

"Let me see those beautiful blue eyes, honey," she would say as he lay there immobile, and sometimes he would smile, as if deep inside he was remembering, or could still feel the warmth of her love.

M. is gone now, his suffering over. But C.'s own health is at risk, worn from the years of battle, fighting to protect her husband.

"Among the top ten causes of death in America, Alzheimer's is the only one which cannot be prevented, cured, or even slowed."[1]

Chances are, you know someone whose family suffers from AD.

Bob Klein's mother passed from Alzheimer's. When I asked him if I could mention that fact in my book, he nodded and said:

"The worst is when they cannot recognize their family."

Six million Americans have the disease of memory loss.

This is not normal getting-old forgetfulness, as when I forget to turn out the light, or lose track of what Gloria is saying; this is forgetting what a light switch is, or who is this person who claims to be my wife?

Eventually the person with AD "forgets" how to chew and swallow, and move and finally, how to breathe. This long slow dying may take ten years or more, during which time they require extensive care.

C. at least was financially stable; she and M. had been high-paid executives.

But for most, Alzheimer's disease is ruinously expensive.

But surely Medicare and Medicaid will provide care?

They do, for a while …

"Medicare will pay up to 100 days of skilled nursing home care under limited circumstances. … long-term nursing home care is not covered."[2]

One hundred *days*? Three months, for a disease that may last ten to twenty years?

So who does provide the needed care? Generally that falls to the family: people who may lose their jobs if they take off too much time to

---

[1] http://www.alz.org/facts/
[2] http://www.alz.org/care/alzheimers-dementia-medicare.asp

care for their loved one. Some of these heroic caregivers are in ill health themselves. This is self-sacrifice on a colossal scale:

"In 2015, more than 15 million caregivers provided an estimated 18.1 billion hours of unpaid care ... AD caregivers "are 28% more likely than other adults to eat less or go hungry because they cannot afford to pay for food ..."[3]

How wide-spread is the problem? According to a 2016 survey:

"Every 66 seconds, someone in the United States develops the disease. Right now, AD kills more people than breast and prostate cancer combined ..."[4]

AD costs our country about $236 billion a year, and worse is on the way.

In the year 2050, there will be an estimated 16 million patients, and the cost may be as high as "one TRILLION dollars a year"[5] (emphasis added) which will almost certainly break our healthcare system.

So, what do we do to prevent what has been called: "the impending Alzheimer's healthcare disaster?"[5]

First, we need to provide research funding on a massive scale. Remember that great line from the movie *Field of Dreams*: "If you build it, they will come?" If funding is provided, the scientists will come.

But right now, research funding is grossly inadequate.

From the NIH, America's largest source of medical research, Alzheimers' disease research gets only about $480 million — less than half a billion dollars.[6]

In addition to vastly increased funding, we need scientists with an unconquerable attitude, who will fight on against this deadly threat, and never quit until it dies.

Fortunately, America has such scientists: people like David Schubert, who for 25 years has pitted himself against Alzheimer's.

---

[3] https://www.ncbi.nlm.nih.gov/pubmed/27570871
[4] http://www.alz.org/mglc/in_my_community_60862.asp
[5] http://www.sandiegouniontribune.com/opinion/commentary/sdut-alzheimers-drug-development-2016jun09-story.html
[6] http://www.alz.org/boomers/

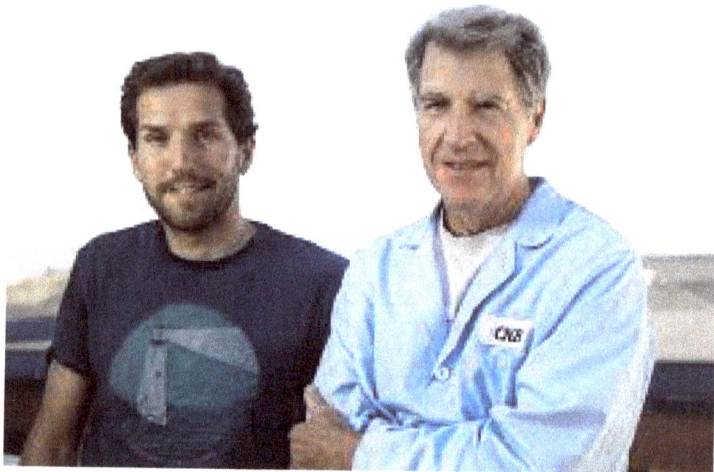

Antonio Currais and David Schubert (salk.edu).

I saw him first at a public meeting of the California stem cell board of directors meeting, March 26, 2015.

Dr. Schubert was there because his request for a grant had been turned down. At this point, most scientists just say, "Oh, well!" and try somewhere else.

Schubert, however, was there to fight: politely of course, but to make his case. He raised valid points, the 29-member board listened to him — and voted. He won the grant, $1.7 million dollars, and so was able to continue on this vital quest.

Consider what he is up against. The brain of a person with Alzheimer's has plaques in it made of a protein, beta amyloid, which kills the nerve cells.

For a therapy to work, the AD brain must be able to do two things: to grow new nerve cells, and protect the old ones.

Schubert came up with a compound, J147, which did both. He, Antonio Currais (who looks like a younger version of Schubert) and other members of their lab at the Salk Institute have been working hard on J147 and other drug candidates.

Interestingly, J147 was originally derived from curcumin, a curry spice much used in India. It had been noticed that India has very little

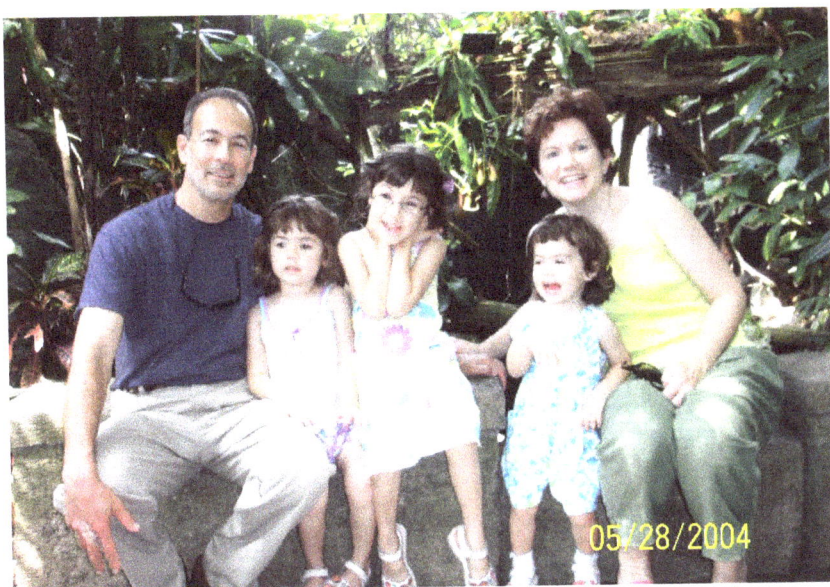

Ernest Villafranca and family (Dr. Villafranca's personal collection.)

Alzheimer's disease. Might their diet be protecting them? Could people just eat some spice and be fine?

No. There had been serious trials to do just that, without success.[7]

But when the scientists tried J147 on a rapidly-aging variety of mice, there was an unexpected side effect.

The treated mice seemed to become <u>physically and mentally younger</u> …

"As (the test mice) aged, those treated with J147 showed improved physiology, memory, and appearance that more closely resembled younger mice … they (also) had better cognition, increased energy, healthier blood vessels in the brain and … a host of unexpected anti-aging effects …"[8]

The latest news on the project?

---

[7] https://alzres.biomedcentral.com/articles/10.1186/alzrt146
[8] https://medicalxpress.com/news/2015-experimental-drug-alzheimer-disease-anti-aging.html

"Abrexa Pharmaceuticals, a San Diego company formed to commercialize J147, has licensed the compound from Salk, and is seeking to raise $10 million to fund Phase I clinical trials ..."[9]

Wanting to know what the clinical trials might be like, I interviewed Dr. Ernest Villafranca of Abrexa. He would be conducting the trials if the $10 million dollars could be raised.

Interestingly, of the first four trials, three would be on people who did not have AD. This was for safety alone, trying to be sure J147 would not itself be harmful. Assuming it proves safe, the 4th trial will be on people with Alzheimer's.

It will be a pill, taken once or twice daily.

Soft-voiced but passionate, Dr. V. had strong views on how to fight Alzheimer's.

"We need to fund research on a massive scale. For instance, the latest federal budget gives an increase to the military of $30 billion. That is an increase, not their total budget. That increase is almost as much as the NIH budget, $32 billion. NIH should be budgeted for at least $150 billion, five times what it is."

Consider the young scientist, Antonio Currais, co-writer of the J147 paper.

"Antonio is doing his post-doctorate work on a grant. What happens to him when that funding runs out? His job at Salk is not permanent; what a crime if such a brilliant scientist could not continue his work!"

Will the 10 million dollars be found in time to fund the clinical trials? I do not know. It is my hope that CIRM will be able to help. If that happens by the completion date of this book (August 15th, 2017) I will include that information. But right now? No idea.

I am only certain of one thing. If we do not **cure** Alzheimer's, we will go broke providing **care**. Unless we abandon our loved ones, care and cure are the only choices we have, and only cure can bring relief.

---

[9]http://www.sandiegouniontribune.com/business/biotech/sdut-alzheimer-j147-mice-brains-2015nov12-story.html

20 PRESIDENT TRUMP'S GREAT STEM CELL OPPORTUNITY

As of this moment, February 13, 2017, I do not know if new President Donald Trump has his mind completely made up on the subject of stem cell research. Certainly I cannot find any quotes from him either for or against the research.

My guess is that — as a person — he will have no automatic animosity against it. And to be the President who helped conquer chronic disease? That would be an accomplishment like winning a world war.

But if President Trump chose representative Andrew Harris (R-MD) as Director of the National Institutes of Health, that would be an irrevocable decision. Siding against stem cell research is not merely to go against a few hundred thousand scientists. The new President would have sided against the one in five Americans who have a disability, and we, their families, who dream of cures.[1]

In California, we voters backed a $3 billion stem cell program.[2]

We felt so strongly about protecting our research that we established protections for it in the California State Constitution.[3]

This was done at least partially because of political ideologues like Representative Andrew Harris, an enemy of embryonic stem cell research since 2005, when he tried to kill a stem cell program for his own home state of Maryland.[4]

---

[1] https://www.census.gov/newsroom/releases/archives/miscellaneous/cb12-134.html
[2] https://www.cirm.ca.gov/patients/stories-hope
[3] https://en.wikipedia.org/wiki/California_Proposition_71_(2004)
[4] http://www.washingtonpost.com/wp-dyn/articles/A21644-2005Mar9.html

Fortunately, he was not successful. But that was not the end of his attacks.

Harris co-sponsored a "personhood" bill, House Resolution 816, which defined every fertilized egg as a person with rights under law. Bear in mind a fertilized egg is essentially liquid, often shed unnoticed in a woman's monthly fertility cycle. Yet personhood bills would criminalize embryonic stem cell research.[5]

President Trump should retain the services of Francis Collins, current Director of the National Institutes of Health. A born-again Christian, Dr. Collins is a moderate, not a boat-rocker. But he is also pro-stem cell, as are the majority of Americans.

The latest major poll asked: "Do you find it morally acceptable to do medical research using stem cells obtained from human embryos?"

Republicans were neutral (50%), Democrats were in strong support, (74%) — and so were Independents (67%).[6]

Dr. Collins is supported by some strong Republicans, including:

1. Lamar Alexander, Chairman, Senate Health, Education, Labor and Pensions Committee;
2. Roy Blunt, Chairman, Senate Appropriations Subcommittee on Labor, Health and Human Services;
3. Fred Upton, Chairman, House Committee on Energy and Commerce; and
4. Tom Cole, Chairman, House Appropriations Subcommittee on Labor, Health and Human Services, Education and Related Agencies.

On December 2, 2016, these Republican leaders sent an open letter to President-elect Trump, recommending the retention of Dr. Collins, saying:

"Dr. Collins is the right person, at the right time, to continue to lead the world's premier biomedical research agency. He possesses all the attributes one should have as the Director of the National Institutes

[5] http://www.ontheissues.org/House/Andy_Harris_Abortion.htm
[6] http://www.gallup.com/poll/170789/new-record-highs-moral-acceptability.aspx

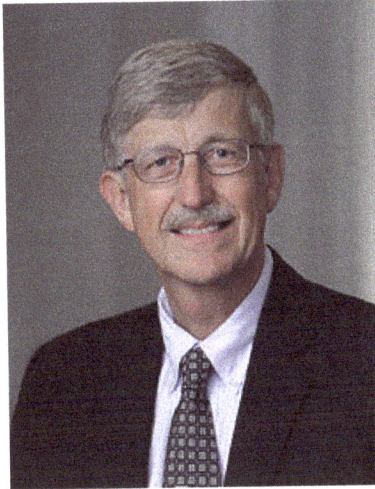

NIH Director Francis Collins (en.Wikipedia).

of Health — intellectual prowess, renowned scientific experience, and outstanding leadership skills. We are confident that under his leadership and with Congress' commitment to biomedical research as a national priority, the National Institutes of Health will thrive and continue to enhance the Nation's health through scientific discovery and biomedical research."

Keeping Francis Collins as NIH Director puts the President on the side of science. But for all who hope for a positive stem cell policy, hiring Andrew Harris would be an absolute deal-breaker.

P.S. I am delighted to report good news. President Trump asked Dr. Collins to stay on as the head of the National Institutes of Health.[7]

---

[7] http://www.medscape.com/viewarticle/874773

21 LEININGEN'S ANTS AND PARKINSON'S DISEASE

Based on the classic short story, *Leiningen's Ants* by Carl Stephenson, the 1954 Chalton Heston movie *The Naked Jungle* is almost unwatchable today: a virtual catalog of racist and sexist attitudes. But it has achieved cult classic standing because of one sequence: the climactic battle with army ants.[1]

A living blanket, the ants invade, overcoming every obstacle. They cross water barriers on floating leaves, sometimes biting onto each other's limbs to make a bridge for the others. (In one ridiculous scene, an overweight native falls asleep on duty, wakes up covered with ants, and is eaten.) The ants are shown as a mindless monster, good for nothing except to be drowned by a conveniently exploding dam.

On Youtube you can see a more realistic depiction of army ants and the people who share their neighborhood — walking calmly beside the dark buzzing column — while the ants eat the pests of the farmers' crops.[2]

What stuck with me was the attitude of the ants. They worked together.

When the army ant (or "driver ant") colony relocates, various ants do different jobs. Some are blind but strong; when an obstacle is in their path, they are brought to the twig or whatever the object, and they lift it, together.

The California stem cell program fights Parkinson's with similar cooperation.

---

[1] https://www.youtube.com/watch?v=qKqninskg74
[2] https://www.youtube.com/watch?v=BVkdw5s3jol

William Langston (www.thepi.org/staff-directory/clinic-providers/j-william-langston/).

Thomas Jovin (www3.mpibpc.mpg.de).

One CIRM-funded project combined the efforts of J. William Langston of the world-renowned Parkinson's Institute, working at long distance with Thomas Jovin of the equally famous Max Planck Institute in Germany.

Xinnan Wang (http://xinnanwanglab.stanford.edu).

The scientists successfully created 51 stem cell lines from skin cells of patients with Parkinson's. Comparing those with stem cell lines of healthy folks might show the weak points of the enemy.

Also needed was a way to attack.

Again, the ant. One species lives in a tree, which it defends against — elephants. When the ponderous pachyderm tries to eat the fruit off the tree, the ants bite the tender insides of the trunk, which drives the huge animal away. Even the most gigantic mammal has a weakness, and the ants exploit it.

Dr. Birgit Schuele has found a way to attack Parkinson's Disease.

Locating the gene which causes PD, (LRRK2.G2019S), Dr. Schuele uses a gene editing tool called the Zinc Finger Nuclease to cut out the bad gene, defeating the disease by altering the human blueprint.[3]

When you watch the ants in action, their energy is astonishing.

In the body, energy is balanced by the microscopic mitochondria. When these stop working, they must be disposed of as waste. Part of

---

[3]https://www.cirm.ca.gov/our-progress/awards/crisprdcas9-mutant-targeting-snca-promoter-downregulation-alpha-synuclein-0

that waste disposal process is a protein called miro. A certain amount of miro is a good thing. But too much?

Stanford's Xinnan Wang's research shows how too much miro can interfere with the body's disposal of mitochondria. So the waste remains, and acts like a poison, killing useful nerves. Does that cause PD? What if the miro was lessened?

"... Prolonged retention of miro ... may (prove to be) a central component of PD ... partial reduction of miro levels ... rescues neuro-degeneration ..."[4]

If you go to the "Parkinson's Page" of the California stem cell program, you will see 26 separate scientific approaches for a total expenditure of $46,606,781.[5] Twenty-six teams, each with a different perspective, but united in a common goal: to defeat Parkinson's.

Oh, and one more thing: when the army ants go to war, they carry back their wounded to the nest: to rest, recuperate, and heal.

Like CIRM which abandons nobody, but fights to bring health to all.

---

[4] https://www.cirm.ca.gov/about-cirm/publications/functional-impairment-miro-degradation-and-mitophagy-shared-feature-familial
[5] https://www.cirm.ca.gov/our-progress/disease-information/parkinsons-disease-fact-sheet

22 ON THE MORALITY OF FETAL CELL RESEARCH

Medical research is organized kindness.

Example: just now I woke up with a sore throat. If it continues, I will go to the doctor, but right now I just grabbed a bottle of throat spray, pushed the aerosol button, aiming the mist down inside my throat, and the pain went away.

Consider: medicines, doctors, hospitals — how many times has your suffering been eased, or your life saved, by a product or procedure made from medical research?

But for scientists to develop cures, or for doctors to administer them, they need cadavers — donated bodies and organs — to learn from.[1]

When my cousin died in a car crash, his mother had to make the wrenching decision to donate his organs. But years later, she met a man who was alive today through the gift of her son's heart. Think of that. Even after his own death, my cousin saved somebody's life.

After I die, I want any part of me still functional to be donated to science.

Organ donation just makes sense. What if someone blind could regain vision from eyes no longer useful to me? Or a liver, or a heart?

Not pleasant thoughts, but important ones: what if organ donation was illegal? Some folks are opposed to anything that alters the body after death; they believe missing organs affect you in the afterlife. I don't agree, but there it is.

Deeper emotional pain applies to a subject few people want to talk about: fetal tissue research. Why does this matter? Think of polio,

[1]https://register.donatelifecalifornia.org/register/

a disease which ravaged the world until Jonas Salk developed the vaccine in 1954. Had polio not been conquered, it would now be costing America more than $100 billion a year. There would need to be actual hotels for people living in iron lungs.

Have you seen an iron lung? A big metal tube encasing your body, so only your head sticks out? A machine makes a vacuum, which noisily helps you breathe, while you lie there, waiting to die.

Instead of that misery, the polio vaccine has saved literally millions of lives, roughly 550,000 people a year. All those families, too, are saved from medical bankruptcy.[2]

But what if the polio vaccine had been against the law?

The Salk vaccine was developed and tested on fetal tissues.

A fetus from an abortion? That's tragic: a pregnancy, terminated, before the fetus was old enough to survive on its own. But if there is an abortion, and I fear there always will be some, should the tissues of the fetus be thrown away, or used to try and save lives?

Done since the 1930s, fetal tissue research has not only wiped out polio, but also eliminated rubella, which used to kill 50,000 babies a year through miscarriages. Fetal tissue was used for vaccines against measles, mumps, chickenpox, whooping cough, tetanus, hepatitis A and rabies, all conditions which once killed people.

Today, fetal cell research is being used to try and defeat the Zika virus, as well as to develop therapies which may eliminate HIV/AIDS. Fetal tissue research might also help end or lessen Early Pregnancy Loss — miscarriages which cost the lives of nearly a million infants every year.[3]

"Since 1994, these vaccines saved society an estimated $1.38 trillion dollars."[4]

Massive regulation controls the obtaining and use of fetal tissue. There can be no profit in its sale (beyond reimbursement of costs), and it can only be authorized once an abortion has been selected. A very clear

---

[2]https://selectpaneldems-energycommerce.house.gov/our-work/benefits-fetal-tissue-research

[3]http://www.hopexchange.com/Statistics.htm

[4]https://selectpaneldems-energycommerce.house.gov/our-work/benefits-fetal-tissue-research

explanation of the various rules was put together by the Congressional Quarterly.[5]

But should we do it? Wisconsin Bioethicist Alta Charo said it best: "We have a duty to use fetal tissue for research and therapy ... Virtually every person in this country has benefited from research using fetal tissue ... every child who's been spared the risks and misery can thank the ... scientists who used such tissue in research yielding the vaccines that protect us ... fetal tissue research (has) saved the lives and health of millions of people."[6]

Unfortunately, when Vice President Mike Pence was governor of Indiana, he signed a law that every aborted fetus must be buried or cremated. This would essentially criminalize the research.

Fortunately, the bill was so extreme a judge found it un-Constitutional, and struck it down, but Pence is still the vice President.[7]

What was Governor Pence's reasoning, when he signed HE 1337 into law? It would, he said: "ensure the final dignified treatment of the unborn."[8]

Dignity after death, is that really what is most important here?

Imagine two children: one is desperately ill, the other is dead. Using cells from one may save the other. If you do nothing, you will have to bury both.

Do we protect the dignity of the dead, or fight for the life of the living?

---

[5] https://fas.org/sgp/crs/misc/R44129.pdf
[6] http://www.nejm.org/doi/full/10.1056/NEJMp1510279#t=article
[7] http://www.indystar.com/story/news/politics/2016/06/30/judge-grants-preliminary-injunction-indiana-abortion-law/86556662/
[8] http://www.vox.com/2016/7/14/12190380/mike-pence-trump-vice-president-abortion-funerals-fetuses

23 DEMOCRACY AND GLORIA'S KNEES

On January 29, 2015, I stood up for public comment at a meeting of the California stem cell program. This is not exactly earth-shattering news. I attend most meetings, and invariably try to sit next to the guest microphone. Unless Gloria is there to restrain me, ("Shh! No more!") I will speak frequently, happily.

But this was different.

I was worried about a change in how money decisions were made. In the past, grants for the scientists had been discussed and decided individually. It was time-consuming, doing them one by one, but it allowed for public input.

Now, the grants would be voted on (by the board of directors, the ICOC) as a bloc: thumbs up or down on all the projects recommended by the advisory board.

This would drastically limit discussion of the projects by scientists, the public, even the ICOC itself.

Bear in mind, the 29-member ICOC is the people's voice. Without its involvement, the CIRM bureaucracy, (50 permanent employees and the out-of-state advisory group), would have unbalanced power. One criticism opponents made was that the advisory group's recommendations were taken too often, making the board a "rubber-stamp."

The advisory group being made of top stem cell scientists from across the nation, it is natural that most of their recommendations will be accepted. As long as the public and the ICOC are part of the discussion — and make the final decision — the "rubber-stamp" charge is not valid.

Jeff Sheehy (blog.cirm.ca.gov).

But without the ability to discuss grants, one by one? The public would be effectually shut out.

True, we would still have the voices of the board-member patient advocates. But the general public, interested outsiders like myself, who might speak up for a particular scientist's grant — we would no longer be part of the process.

Now I love the CIRM. I feel about it the way Biblical Ruth felt about her husband: "Your friends shall be my friends, and your enemies, my enemies."

But should I remain silent, and watch a mistake become permanent?

Here is a distilled version of my comments, from the official transcript:

"We have a saying in the disability community: "Nothing about us, without us." If decisions concerning us are made without our input, that is like patting us on the head and sending us out of the room while the grownups decide …"

"… (This would mean) denying … the help of your strongest supporters and greatest friends. Who built the CIRM? Patient advocates. Who

Jonathan Thomas (Cirm.ca.gov).

raised the money, gathered the signatures, sacrificed our personal lives volunteering for a nearly two-year campaign? Without patient advocates, none of this would be here. And if there comes a time when further funding is required, it will be patient advocates fighting for it again ..."

"Do not exclude us ... Give each of us our three minutes to speak before a decision is made. The board can decide if there is merit in our arguments."

"I formally request that this matter be considered, and, if approved, become a matter of standard policy. Thank you."[1]

The matter was "taken under advisement," meaning they would think about it.

At the very next meeting, March 26th, 2015, Jeff Sheehy and CIRM attorney James Harrison walked us through a "new wrinkle in ... the process."

---

[1] https://www.cirm.ca.gov/sites/default/files/files/agenda/transcripts/ICOC-1-29-2015%20Transcript.pdf

As CIRM Attorney James Harrison put it, in one key sentence:

"Members of the public will continue to have the opportunity to offer public comments before any individual vote is taken."

The Chair of the ICOC, Jonathan Thomas, made it crystal clear.

"... we plan to offer members of the public the opportunity to comment about any application immediately after the CIRM team makes its presentation regarding the Grant Working Group recommendations and before the Board considers any motions to move applications from one tier to another ..."

"This will allow members of the public, including patient advocates, to provide input at the beginning of the Board's consideration of the GWG's recommendations and before the vote on each motion."

— Jonathan Thomas, Chair, California stem cell board of Directors.

Jeff Sheehy took the mike.

"So is that well understood? Are there any members of the public who would like to address any of the grants that we have coming up?"

The first to take advantage of the new policy? Not hard to guess! I had read the agenda (always posted online, before ICOC meetings) and supported one particular proposal: to use stem cells to regenerate cartilage and bone in an arthritic patient's joints. It would not affect Gloria, being aimed at younger sufferers. But still it was about arthritis — and it was presently going to be denied funding ...

I told the board about Gloria's knees:

"... I live in a two-story house ... Gloria has arthritis in both knees, and tries to plan her days so she only goes down the stairs once in the morning and back up at night. The cartilage cushion in her knee joints is worn out, so the knee grinds bone on bone ... Arthritis may affect as many as 18% of all women over 60 ...

"Grant number PC1-08128 looks like a winner. It reminds me of the ... highly successful project by Dr. Sue Kimber of the University of Manchester in the UK. Precursor cartilage cells were implanted in (the) damaged cartilage of rats ... After 12 weeks, the cartilage was smooth and similar in appearance to normal cartilage. Substantive progress toward curing arthritis would be a home run for (CIRM)."

The very next person in line said:

"My name is Darryl D'Lima. I am the PI (Principal Investigator) on the same grant that Mr. Reed spoke about. In the interest of full disclosure, I don't know Mr. Reed and I haven't solicited his opinion ..."

He continued:

"The annual cost of treating arthritis in the US is estimated to be over $200 billion ... Over a million joint replacements are performed in the US alone (each year). However, for younger patients with severe arthritis or impending arthritis, there is as yet no treatment ... Our approach is ... constructing scaffolds that are seeded with (stem) cells programmed to (become) bone and cartilage cells ..."

Dr. D'Lima used his three minutes to answer objections to his project, concluding by saying:

"We know the existing cell therapies don't work. The people who know the unmet needs are the patients, and the physicians who treat them. The data ... is very compelling. ... We've tested against adult stem cells, we've tested against bone marrow stem cells, and we've tested against ... microfracture (a series of small holes in the bone, which stimulates bone marrow growth and some healing — dr), which some

Darryl D'Lima (Scripps.org).

claim is the standard of care, and (our approach) is far better than all three of those ..."

When the vote was taken, Dr. D'Lima won his grant.[2]

Afterward, I made it a point to telephone him. Dr. D'Lima had been a surgeon at first, but found that let him only help people individually, while a research doctor's findings can benefit many.

He chose to fight arthritis, because: "Cancer steals your life. But the pain of arthritis steals your soul."

Why youth-onset arthritis, instead of the old-age variety?

"With youth-onset arthritis," he said, "A person can limp along in extreme pain for twenty years before the arthritis is "bad enough" to justify reconstructive surgery."

He did not want to rely on old methods forever, saws, hammers and drills, but preferred a biomedical approach.

"When the metal version goes into the knee, that is as good as it gets. But with a life-sciences approach, we hope for continual improvement: a self-healing knee ..."

And the California stem cell program? It remains fully interactive with the public. Anyone who wants to speak will have three minutes, guaranteed. Sometimes you will win your point, sometimes you may not, but you are always welcome.

And you will be heard.

---

[2] https://www.cirm.ca.gov/sites/default/files/files/agenda/transcripts/ICOC-March%202015_Transcript.pdf

24 THREE CHILDREN, AND THE ETERNAL FLAME

The boy's name I do not know. (I found his story in the New England Journal of Medicine, volume 344, No. 23.) His condition was called IPEX, which is actually a whole series of diseases, so complex it takes its name from the symptoms:

"Immune dysregulation, Polyendocrinopathy, Enteropathy, X-linked, problems with the immune system, endocrine glands, and intestine...

"Untreated, IPEX syndrome is usually fatal during the first year of life ..."[1]

His life was a long series of hospitalizations. At four weeks, persistent diarrhea ... (causing weight loss) and then anemia; at two months he was found to have type-one diabetes. Hepatitis followed. IPEX was "suspected and confirmed."

At four months, he was given a bone marrow transplant from his 18-year-old sister. The doctors did everything they knew. Their efforts helped; they gained time ...

But not enough. At age 3, the boy whose name I never knew slipped quietly away.

Today, at Stanford University, Dr. Rosa Baccheta continues her 20-year effort to defeat this complicated and deadly disease. But it takes funding.

In her grant-request letter to the California stem cell program, she said:

---

[1] http://www.merckmanuals.com/home/hormonal-and-metabolic-disorders/polyglandular-deficiency-syndromes/ipex-syndrome

Rosa Bacchetta (med.stanford.edu).

"IPEX is a prototype of a series of diseases ... that severely affect children at a very early age ... although this is a rare disease, (curing) it could potentially benefit people with other similar diseases."[2]

Her grant was unanimously approved.

The second child?

Gwendolyn Strong was seven years old, the daughter of Bill and Victoria Strong. Her age was a silent testament to their dedication and love. Most children with Spinal Muscular Atrophy, (SMA), die at the age of two, or younger.

SMA is a form of paralysis, which is invariably fatal.

Roman loved to visit Gwendolyn, and would sit by her home hospital bed and talk to her by the hour. Her huge blue eyes glued on him, Roman would tell her about stem cells, and his children; he felt he could understand what she was thinking.

Lately though, her parents said in their blog that she was tired, and her anxiety level was high. So was theirs. They stayed by her constantly, day and night.

When a child has SMA, his or her body loses so many abilities, like swallowing. Gwendolyn had to be fed by a tube in her stomach. She

---

[2]https://www.cirm.ca.gov/sites/default/files/files/agenda/CIRM-written%20comment%20to%20ICOC_DISC2-09526.pdf

Gwendolyn Strong and her parents (http://thegsf.org/).

had never known the joy of chewing a good solid meal. She could not run, or walk, or move anything under her own control. She could not even breathe for long, without the machine. It was frightening when the oxygen tank which gave her breath had to be replaced.

She was beautiful: her eyes incredibly blue, like windows to the sky.

Gwendolyn's last name was perfect for her: Strong. Her parents, Bill and Victoria, raised money for research through the Gwendolyn Strong Foundation.

Champion scientists like Hans Keirstead worked to develop new therapies, trying to defeat this vile killer of children.

When Roman called me on the phone, he was so racked with sobs I could not understand him at first, when he tried to tell me Gwendolyn had died. She was the same age as my grand-daughter, Katherine.

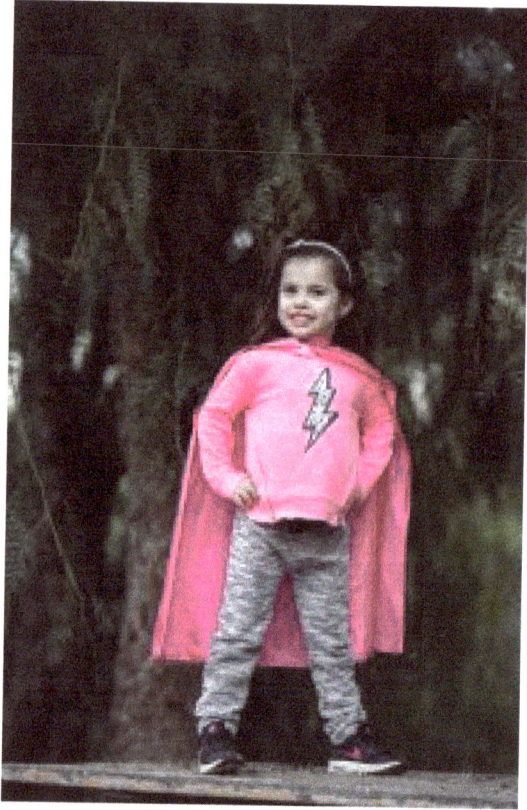

Evangelina Padilla-Vaccaro (blog.cirm.ca.gov).

The Strongs fight on, so others will not have to endure what they all did.[3]

The third child was an infant, Evangelina Padilla-Vaccaro. She had an immune system disease: Severe Combined Immune Disease (SCID), better known as "bubble baby" syndrome. Made famous by a John Travolta movie, "Boy in the Plastic Bubble," the condition often kills in the first few months of life.

"I knew something was wrong right away," said her mother, Alyssa Padilla-Vaccaro in a telephone interview: "She had a gray color, did not

[3]http://thegsf.org/donate

Donald Kohn (uclahealth.org).

move right, and she spit up a lot. ... SCID meant she had no immune system; her body had no way to fight back against germs; a cold could kill her. We made our house as sterile as we could. My husband Christian and I wore masks all the time. Evangelina and her twin sister Annabelle never saw our mouths.

"Then I heard about stem cells and UCLA and Dr. Donald Kohn, and how it might be possible to rebuild Evangelina's immune system.

"But stem cells? I had been raised a good Catholic girl, and had heard all kinds of bad things about stem cells. But when it is your child in danger, you will listen very carefully, and we did. This was about saving lives.

"There was just one opening left on the clinical trials ... Dr. Kohn held it open for the two months it took our daughter to grow strong enough to endure the operation.

"Dr. Kohn was up front with us ... never promising a cure. But my husband and I had read more than 30 studies on the National Institutes of Health website, trying to educate ourselves on what had been tried before on our daughter's condition, and Dr. Kohn's approach made sense.

"He would take out bone marrow from her hip. There was a mutation in her genes. He would fix the gene, put it in some stem cells, and put it back.

"I had had three miscarriages before the twins arrived ... was I now to lose another of my children?"

There were 19 children with SCID in that clinical trial. One boy died, from an infection he had had earlier.

But the other 18? They are alive and well to this day.

I met Evangelina and her family at a California stem cell board meeting.

Evangelina spoke. She was four years old, spunky, wearing a pink superheroine T-shirt. She was startled by the squeal of microphone feedback, but stuck to her task.

"Thank you," she said, in a tiny voice.

Her Mom also spoke to the CIRM: "Thank you, for keeping my family together."

When I talked to her later, she added, "If there is an effort to build Prop 71 part two, count me in. I will do anything I can to help."

One commentator, who shall be nameless, said California had spent millions of dollars just to save a few children.

Kevin McCormack, spokesman for CIRM, responded:

"Some might ask: why spend limited resources on something that affects so few people? But there is no hierarchy of need. We are not in the business of making value judgments about who has the greatest need. We are in the business of accelerating treatments to patients ... and the lessons we learn in developing treatments for a rare disease can often lead us to treatments for diseases that affect many millions of people ..."[4]

How can a scientist keep struggling to find a remedy for disease, even knowing cure may not happen in their lifetime?

I do not know. But when I think of scientists like Kohn and Keirstead and Baccheta, I remember visiting the Tomb of the Unknown Soldier, late one night in Washington DC.

A small flame flickered in a sheltered recess, in the marble monument: just an orange flame wavering in the night breeze: but it did not go out.

That monument honors soldiers who sacrifice their lives in the service of our country. They deserve our eternal respect.

And so do our medical researchers, who fight for the lives of all children.

---

[4]https://blog.cirm.ca.gov/2017/02/28/raising-awareness-about-rare-disease-day/

25 AUTISM, MINI-BRAINS, AND THE ZIKA VIRUS

Alysson Muotri will probably always be remembered for his "Tooth Fairy" methodology, a way to get tissue samples from children with autism.

Consider the difficulty. Many kids with Autism do not like being touched. When my wife Gloria was a Special Education teacher's aide, she spontaneously hugged one autistic student. He screamed so loud, they both fell off their chairs!

Now, imagine taking a skin sample (biopsy) from an autistic person...

Easy! If you use the "Tooth Fairy" method ...

You know how it works. The child places his or her baby tooth under the pillow; the Tooth Fairy trades for it, leaving a little bit of money. When I was a kid, a dime was left. Then inflation raised the exchange to a quarter, and now I don't know what the going price is — probably stocks and bonds!

But from the fresh pulp inside a newly-lost tooth, Dr. Muotri can make stem cells: and from them a miniature "brain," tiny as the dot of ink above this letter "i." That speck of living tissue, a mini-brain ("cerebral organoid") can be used to test drugs to find the cause of autism.

For instance, does living near a highway increase your chances of autism? Exposing a minibrain to auto exhaust might let us know.

Dr. Muotri has a ten year old stepson, Ivan, who has Autism. I hoped he would not mind if I asked him a personal question about this.

An opinion had been voiced that it was actually a "privilege" to have Autism, implying that to try to remove the condition was unnecessary, perhaps even wrong.

Alysson Muotri (ucsdnews.ucsd.edu).

"Autism is a spectrum," Dr. Muotri said, "from mild to severe. Some who are lightly touched by the condition can live a relatively normal life. They can marry, hold down a job, live independently. But my son is non-verbal. He cannot speak. Some things other children learn at a much earlier age he cannot do at present, and possibly never will … He has not yet mastered brushing his teeth or combing his hair. On top of that, he now has seizures."

"Ivan requires one-on-one supervision. My wife Andrea quit her job so she could stay home and take care of him all day. I take over during the weekend so she can have a break. But even so, Andrea had to subtract herself from the workforce, and set aside so many of her hopes and dreams …"

The voice on the phone went silent. I felt a need to change the subject.

There was another reason I had been eager to catch up with Dr. Muotri.

Zika.

The Zika virus is transmitted by mosquitoes, and has spread from Africa and Asia to South America. It affects unborn children, who may have under-developed brains, even physically smaller heads.

But the newer variety of Zika, from Brazil — does this strain also cause birth defects? Unfortunately, Dr. Muotri's work established that it does.

Can anything good come from knowing that?

Yes. If we are to defeat a disease, we must know its cause, and not be distracted by false issues. Wherever there is such a dramatic health issue, people try desperately to find the cause. This is only natural, but it can take us off the track.

"For instance, it was once mistakenly believed that autism might be caused by childhood vaccinations. That has been proven completely wrong," he said.

Knowing that the Zika virus was the cause of the micro-cephalic (small-brained) babies in Brazil was vital: allowing us to attack the right enemy.

Do you remember the measles epidemic? Children died by the hundreds, were hospitalized by thousands, and suffered by millions. A rash might spread over the entire body, itching so severe a patient's hands must be tied, or they would scratch themselves bloody. And the worst side effect? Encephalitis, swelling of the brain.

Dr. Muotri worries that the Zika virus may begin another wave of autism. His data suggests that Zika- infected babies, even those born with normal-appearing heads, may incur developmental delays and serious cognitive problems.

But something else is possible: something wonderful.

On a grant from the California stem cell program, Alysson Muotri may have found a way to block the transmission of Zika from mother to her unborn child.

Using human minibrains in the lab, (created through those baby tooth stem cells), Muotri tested drugs to inhibit the viral replications in the brain. He found one FDA-approved drug that worked really well: designed for the hepatitis C virus.

His lab team tested the drug on mice. It proved to be effective in adult animals, reducing the viral load to undetectable levels. They also tested pregnant mice. In this case, the "moms" were previously infected by Zika and then treated. The baby mice, though born from infected moms, showed no evidence of the Zika virus.

If the effect of the drug is similar in humans, it would be the first successful treatment of the Zika virus.

A gift from the Tooth Fairy that could change the world.

26 WHY "THE BIG BANG THEORY" MATTERS TO ME

As it is for millions, my favorite TV show is *The Big Bang Theory* (TBBT), written and produced by Chuck Lorre, Bill Prady and Steven Molaro.

Everything works. Penny (Kaley Cuoco) is not only impossibly beautiful but also one of the most talented comediennes in entertainment history. Sheldon (Jim Parsons) is pure gold as the "wise fool" who speaks his truth regardless who it hurts. Leonard (Johnny Galecki) is utterly lovable as the nerd who somehow wins fair maiden — all of the actors are excellent: there are no minor players.

The show is warm, inclusive, hilarious, and something else.

TBBT says that science is not to be feared.

Remember the ending of the 1931 classic movie *Frankenstein*? The villagers came with pitchforks and torches, to kill the scientist and the monster he created. Boris Karloff played the monster and later the "mad scientist" all his career.

"He meddled in things man was meant to leave alone," (from *The Invisible Man* with Claude Rains, 1933) sums up the anti-science attitude.

That mindset continues to this day, and unfortunately not just in the movies. I am sick and tired of hearing politicians say, "I am not a scientist!" to avoid taking a position on global climate change.

We all need to know at least a little science, to protect our children's future.

Strangely, I did not watch "The Big Bang Theory" for years, though Roman and Gloria kept telling me, "Watch this show, Sheldon is just like you!"

I am not like Sheldon, I explained to them — he is a physicist.

THE BIG BANG THEORY (www.the-big-bang-theory.com).

But I do understand Sheldon's answer to the question:

"What was it like, growing up in East Texas?"

"It was Hell," said Sheldon.

As a kid I combined physical frailty (asthma like Leonard), a big mouth like Sheldon, and an occasionally problematic IQ level. I was surprised to learn that people did not appreciate being corrected. My religiously conservative parents were not eager for my questions on creation; my teacher objected to reading books about sharks during math; the schoolyard was a place of terror and frustration.

Being bullied pushed me to learn weightlifting, to defend myself. I became strong, punched out my nemesis, and later competed in Olympic-style weightlifting.

But I never really lost the feeling that intelligence was something best kept hidden.

So when Gloria turned me on to TBBT, I became an instant fan.

And smiling down from my wall today are autographed pictures of three stars from the show: Kaley Cuoco, Johnny Galecki and Mayim Bialik.

The heart of the show? Being smarter than average can be difficult (and hilarious), but it is not a crime. Scientists share the same problems that trouble us all.

Six of the characters have jobs in science. They struggle with actual physics difficulties. Real scientists like the great Stephen Hawking visit the show and share the laughter; a working physicist, David Salzberg, checks the scripts (and the white boards on set) to make sure the scribbled diagrams are correct.

They even mention stem cells!

So valuable … The world needs more real-life people like Sheldon and Leonard and Bernadette (Melissa Rauch) and Amy Farrah-Fowler (Mayim Bialik, in a real life a PhD neuroscientist); they fight giant problems which threaten everyone.

One show depicted Sheldon, Howard, and Leonard's attempt to encourage young women to consider science as a career. When they fail, Rauch and Bialik come to the rescue, although they and Cuoco are dressed as Disney princesses.

No "glass ceiling" should block female scientists, or we deny ourselves half the world's mental contributions; it does not take big arms to pick up a test tube.

But it works both ways. Scientists (if they want public funding) must develop their "people talk" skills. Like a race car mechanic, they possess specific knowledge and a difficult vocabulary. They should listen to the radio show of "Klick and Klack, the Tappet Brothers," who make auto mechanics interesting and fun.

Scientists need small words and vivid examples, as when Leslie Winkle (played by Sara Gilbert) demonstrates extreme cold by flash-freezing a banana in liquid nitrogen, and shattering it to put the fragments on her cereal.

In "Bang," science is not presented as dark and mysterious. The work they do is hard mental labor, like figuring out blue prints, or tax returns. It requires practice, but is doable. The basics can (and should) be understood by everyone, so we will not be cheated.

For instance, evidence-based reasoning, science can be trusted because it depends on proof. Every assertion must be backed up by a repeatable experiment. If something cannot be repeated, it probably is not true.

If we discredit science, or act in ignorance of it, we risk disasters, like a new Dark Age, when preventable diseases wiped out a third of Europe.

Be glad for those who love science, the way a chef loves food we enjoy; or an architect prevents a skyscraper from collapsing: they are all involved in mystery, from which we benefit.

And speaking of mystery, would it not be great for Jim Parsons and Johnny Galecki to play Sherlock Holmes and Dr. Watson? And of course who would be better than Kaley Cuoco as Irene Adler, the woman who outwits them both, and wins the heart of Sherlock?

Hats off to everyone connected to *The Big Bang Theory*! You bring happiness to the world.

May your careers be long and varied, and your repeats play for a thousand years.

27 MUSASHI AND THE TWO-SWORD SOLUTION

The greatest real-life swordsman in history, Musashi Miyamoto* (1584–1645) fought 60 battles and never lost. The latter assertion may be judged by the fact that he died of old age; most sword fights were to the death. The fact that he was still standing afterward meant he did not have his skull crushed or his heart pierced.

Musashi took his place in history by a literary contribution, *The Book of Five Roses*, which he wrote in a cave during his last years, and which is respected to this day in business, martial arts, and philosophy circles.

His soldier father, Muni, was an ill-tempered brute. One day he threw a knife at 8-year-old Musashi's head, and became furious when the boy ducked. He had only meant to nick his ear to teach him manners; he should have held still!

Musashi walked out of the house, and went to live in the forest.

At 13 he had his first formal duel. The rival was a full-time swordsman, who put up a sign, challenging everyone. Musashi scribbled his acceptance on the sign. He had no sword, but just walked up to the soldier and killed him with a wooden pole.

Amazing reflexes: he once placed a grain of rice on his adopted son's head, swung his sword and cut the rice, but not the son. He then repeated it four times.

---

*The most accurate book about real-life Musashi is probably "Miyamoto Musashi: A Life in Arms," by William De Lange. The most exciting book is the novel "Musashi," by Yoshikawa Eiji, often called the Japanese "Gone with the Wind." A movie trilogy, "Samurai," starring Toshiro Mifune and directed by Hiroshi Inagaki, earned an Academy Award as best foreign film in 1954.

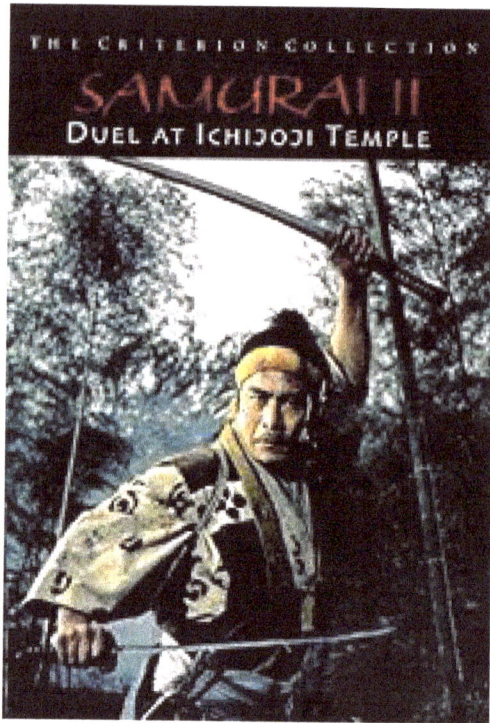

Musashi Miyamoto as portrayed by Toshiro Mifune (trailers.apple.com).

At one point, Musashi literally fought an army. He defeated Seijuro Yoshioka, leader of a famed martial arts family, by a trick. Pretending to be sick, Musashi was carried to meet his opponent. He jumped out of the borrowed palanquin, knocked his enemy unconscious with one blow of a home-made wooden sword, then calmly walked away. (Seijuro recovered fully, and became a monk.)

But Seijuro's brother Denshichiro was furious at the perceived insult to the family's honor. He challenged Musashi, coming at him with a five foot sword. Musashi yanked it away, and killed Denshichiro with his own weapon.

The Yoshioka clan was wild for revenge. A challenge was issued by Denshichiro's teenaged son, Matashichiro. He was only 13, the same age as Musashi has been at his first duel, but there was no intention of having him fight. It was an ambush. All the fighters in the clan came together, supposedly for a practice session, in great numbers, historical estimates running from a low of fifty to as many as "hundreds." They thought it would be a simple assassination.

Instead, Musashi leaped from a tree, shouting, "Did I make you wait too long?" He cut down Matashichiro with a blow, and then attacked everyone: herding, driving, killing. Most ran. Those who stayed, died, their bodies strewn like leaves after a storm. The Yoshioka clan never recovered, broken by a single man.

Musashi was more peaceful in old age, but would still fight anyone who wanted it. He did not kill them (usually): merely chasing them backwards until they got tired.

How was Musashi able to defeat everyone? He had astonishing athletic ability of course, plus endless training and outdoor life, and an incredibly violent lifestyle.

But there is perhaps another reason: a secret in plain sight. In those days, a samurai carried two swords: a long one for battle, and a short one for seppuku, ritual suicide, if its' owner could not live with honor. Soldiers practiced hours every day, perfecting their skills with one weapon.

Musashi fought with both swords at the same time and was invincible.

And now we come to a modern-day battle facing every stem cell researcher. It has a fanciful name, the "Valley of Death," but is very real; scientists who cannot fight their way through it are doomed to watch their research die.

The Valley of Death is the FDA regulatory process, that seemingly endless series of tests, trials, and requirements, to be sure all new drugs and procedures are safe and effective. The FDA is necessary of course: for safety, and to prevent fraud.

But still the process is time-consuming, and expensive. It may cost (literally) billions of dollars and decades of delay before a new medicine is released.

Remember the stem cell therapy begun by Roman's Law? Correspondence with the FDA exceeded 22,000 pages, and the testing has taken over 15 years so far!

The greatest stem cell theory is of no value, until it is put into practice. Would Edison's ideas on electricity have mattered, if they had not led to the light bulb? We need a way to help translate the scientists' ideas to everyday reality, and to accelerate that process.

CIRM President Randy Mills told California Stem Cell Report's David Jensen:

"Many scientists are brilliant researchers but have little experience ... navigating the regulatory process. (But) they don't have to develop those skills, we can provide them ..."[1]

For instance, before a scientist can officially begin the FDA approval process, he/she needs official permission, the coveted IND (Investigational New Drug) approval. How can they best prepare for this? And what about the series of tests for safety and efficacy? Can scientists and an understaffed FDA work together?

To meet this challenge, the California stem cell program is developing a Stem Cell Center, to help translate theories into therapies. A $30 million project, the Stem Cell Center will be run by Quintiles Transnational.

"Quintiles ... is world-class when it comes to regulatory affairs and consulting and clinical trials[2] ..." said Joe Panetta, ICOC board member, and President/CEO of Biocom, a life science trade group.

This may benefit the entire field of medical testing. If we can make the approval process more efficient, we may lower the costs of all new medicines.

With funding from the California program, and guidance from the Alpha Sites and the new Stem Cell Center, our scientists will be fighting with multiple swords!

Like modern-day Musashis for medical research ...

---

[1] https://www.cirm.ca.gov/about-cirm/newsroom/press-releases/06152016/cirm-creates-first-its-kind-center-accelerate-stem-cell

[2] http://www.huffingtonpost.com/don-c-reed/of-starfish-stem-cells-an_b_3473674.html

28 "THE MAGNIFICENT SEVEN"

Remember the classic Western, *The Magnificent Seven*, starring Yul Brynner, Steve McQueen, Charles Bronson, and Eli Wallach? This was of course based on "Seven Samurai," by Akira Kurosawa, starring Toshiro Mifune.

The story was about a poor village which gambled everything they had — a gold watch and a little cash — to hire gunfighters to drive the bandits away. The fighters came, each with a different set of deadly skills; working together, gunmen and farmers saved the day.

But imagine the story, without the gunfighters — just the bandits and the village.

Who wins? Does not the suffering just continue?

Now. Does your state have a stem cell program, scientists fighting to cure disease?

Probably not.

Across America, only seven states currently fund stem cell research programs: California, Texas, New York, Maryland, Minnesota, Washington, and Connecticut.

Illinois, New Jersey and Massachusetts used to have stem cell programs, but the money ran out and their programs, to the best of my knowledge, were not renewed.

Even in my beloved California, danger is in sight. At our present rate of funding research, we can go on until about 2020, and then no more. Granted, our Chairman Jonathan Thomas is an expert fundraiser. If thing get too tight financially, we can count on him to hustle up some funds.

But we don't need "some" funds; we need a lot! Every state in the union, indeed every country in the world, is up against a veritable plague of chronic diseases.

Some people say, let the Feds take care of it. Let the National Institutes of Health (NIH) be the sole source of funding. To me, that is like saying, we have the Army, so we don't need local cops, sheriffs, or the Highway Patrol!

That is why this book exists: to encourage every state to fund research, and most especially California, because there absolutely must be a Proposition 71, Part 2.

More on that later!

And the other six states? Let's take a look.

TEXAS has the Cancer Prevention Research Institute of Texas (CPRIT), a $3 billion program developed by champion bicyclist/patient advocate Lance Armstrong. In many ways, it is a great program. If I were a young scientist interested in basic research for cancer prevention, or to study the mechanisms of how cancer works, Texas is where I would go.[1]

But to actually defeat cancer, to destroy it for good and all?

For that we need the full toolkit of stem cell research, and the Texas program is not that. CPRIT does fund some stem cell projects, but none of it is embryonic. That is still legal in Texas, thanks to the heroic efforts of advocates like Nina and Joe Brown, Judy Haley, Beckie McCleery and their outfit TAMR (Texans for the Advancement of Medical Research), but no actual money gets spent on it. Politicians (almost exclusively Republicans) block it.

As Texas Cures advocate David Bales said in a personal communication, "When you bring up the subject of embryonic stem cell research, the politicians just say 'This conversation is over,'" and then they leave the room."

David is a real powerhouse, whose specialty is the protection of patients, so that they will not be ripped off by charlatans promising cures they cannot deliver. Look him up in "Right to Try" battles, where the FDA may be denied involvement with experimental medicine, and patients pay to be included in clinical trials, but cannot sue if something goes wrong. "Right to Try" sounds good, but contains danger.

Perhaps in time Texas will expand its program to include full stem cell research and a reliable oversight system; but that time is not now.

[1] http://www.cprit.state.tx.us/

MINNESOTA owns the newest state-funded stem cell program in the country, Regenerative Medicine Minnesota. It began in 2014, to provide $4.3 million a year for ten years. It is administered through a partnership between the University of Minnesota and the Mayo Clinic, and has four funding goals. They are building the state's regenerative medicine infrastructure through research, biobusiness development, creation and retention of a trained workforce, and eventually improved patient access to the new therapies.

Developed primarily through the efforts of Dr. Jakub Tolar and legislators like Representative Erin Murphy, the program emphasizes attracting new blood to the field by focusing research funds on young investigators. Dr. Tolar was a grant reviewer for CIRM and has enormous enthusiasm for it.[2]

CONNECTICUT: Sadly, Connecticut's program is at risk.* Its $10 million a year is spread among the University of Connecticut, Yale University and Wesleyan University. A tremendous success, even in financial terms, $94 million spent has brought an astonishing 6-1 return, more than $600 million in add-on grants, new money from the National Institutes of Health and other sources.[3]

Note: For more complete information on Connecticut and several of the other states mentioned in this chapter, see my first book: *Stem Cell Battles*.

NEW YORK: the Empire State has two strong programs, one private, one public. Susan Solomon, CEO of the New York Stem Cell Foundation, raises private money (about $20 million a year) for stem cell research.[4] The New York Stem Cell Research Program (NYSTEM) is public funding, about $50 million a year.[5]

---

[2] http://www.regenmedmn.org/

* As of today, March 11, 2017, Connecticut's program is being denied its funding. Although $6 million was already allocated this year, that money may be held back, as well as the next two years' funding of $20 million. It is to be hoped that the citizenry of Connecticut and every state will see their way clear to the continued support of regenerative medicine, for the good of all.

[3] http://ctinnovations.com/rmrf

[4] https://www.nyscf.org/our-research

[5] https://stemcell.ny.gov/

David Bales (texansforcures.org).

Ren He Xu and Xiaofang Wang (tina encarnacion/uconn health center photo).

MARYLAND has an $8 million annual program, the Maryland Stem Cell Research Fund, MSCRF, with what appears to be just two years left in the program. So far, it has funded over 375 stem cell projects, investing more than $120 million in Maryland stem cell research. The Executive Director is Dan Gincel, Ph.D. Grants run from $110,000 to $750,000 over a maximum period of three years. Gincel cites "Insufficient funding at the federal level for all types of … human stem cell research" as the reason for the program's inception.[6]

WASHINGTON's program is called "Ice Cream" by insiders, an acronym for their Institute of **S**tem **C**ell **R**esearch and **R**egenerative **M**edicine (ISCRRM).

ISCRRM is currently asking for a state investment of $3 million a year to "support our mission to turn fundamental discoveries in stem cell biology and tissue regeneration into lifesaving medicine. The requested funds will allow ISCRRM to recruit and retain the best faculty, staff, students and trainees, to perform the highest quality research and training, and conduct key clinical studies that will establish Washington's medical centers and biotech companies as an epicenter for regenerative medicine." — U.W. Institute for Stem Cell and Regenerative Medicine, internal document.

ISCRRM is led by two highly qualified faculty members: "Director Charles Murry, Professor of Pathology, Bioengineering and Medicine/Cardiology, and co-Director C. Anthony Blau, Professor of Medicine/Hematology." (Dr. Randy Moon was unfortunately forced to retire, by Parkinson's Disease.)

Dr. Murry's lab has been "investigating stem cell-based approaches to heart repair for nearly twenty years." Martin (Casey) Childers' love of dogs, especially a Labrador retriever, "Nibs," has led to a focus on the deadly disease of Muscular Dystrophy, which attacks both people and dogs.

Pamela Becker is seeking to utilize "Precision Medicine for Acute Leukemia"… Michael Regnier has shown that by using an enzyme, ribonucleotide reductase, it is "possible to increase contractile force of heart muscle cells," which may become extremely important in heart repair …

---

[6]http://www.mscrf.org/

Only seven state programs? Where is Missouri?[7] Or Michigan?[8]

In both states, patient advocates fought and won tremendous battles for scientific freedom. Proposal 2 and Amendment 2 were enormous and exhausting victories. I helped out in both efforts, and it was a joy to watch patient advocates unite and fight. The anti-research forces flooded the airwaves with grossly deceptive commercials. But the voters saw through the nonsense, and legalized the research.

But our victories were hollow, because anti-research ideologues remained in power, and blocked the funding. Nothing happened after that, except frustration.

We need patient advocates and scientists, working together for funding and freedom, or our people will go on dying, like villagers and bandits.

---

[7] https://ballotpedia.org/Missouri_Stem_Cell_Research,_Amendment_2_(2006)
[8] https://ballotpedia.org/Michigan_Stem_Cell_Amendment,_Proposal_2_(2008)

29 THE CONNECTICUT COMMITMENT

What is so powerful a goal that a man would die for it? Hold that thought.

In 2005, then-Representative Chris Murphy authored and helped pass the $100 million Connecticut Stem Cell Investment Act. Signed into law by Governor Jody Rell, the program later became the Connecticut Regenerative Medicine Research Fund, continued under Governor Dannel Malloy.

It worked. As (now U.S. Senator) Murphy put it, "We have over 100 labs … almost 200 researchers … doing embryonic and related stem cell research — it's an amazing, amazing thing."[1]

Governor Malloy stood by the program. A forward-looking man, Malloy has been a featured speaker at the International StemConn meeting for years. I first heard him speak at the World Stem Cell Research Summit in 2009, and remember thinking: presidential timber.

Connecticut, a small state, has achieved world renown in stem cell research. Why? Top-flight educational institutions collaborated: Yale, the University of Connecticut, Wesleyan University and most recently Jackson Laboratory.

Example: Epilepsy is characterized by "the loss of interneurons, leading to over-excitation in the brain and therefore seizures." Connecticut scientists Drs. Janice Naegele, Gloster Aron, and Laura Grabel are testing stem cell therapies for Temporal Lobe Epilepsy, to see if embryonic stem cell-derived interneurons can suppress seizures …" — personal communication, Laura Grabel.

[1] http://www.huffingtonpost.com/don-c-reed/racing-for-senate-in-a-st_b_1664334.html

Laura Grabel (http://lgrabel.research.wesleyan.edu/laura-grabel/).

Milton B. Wallack, pioneering Connecticut stem cell advocate (jewishledger.com).

Example: Dr. Caroline Dealy is working on a cure for arthritis: to replace worn-out cartilage with specialized cells called chondrocytes.[2]

Example: Drs. Stormy Chamberlain and Marc Lalande are working with stem cells derived from patient skin cells to try and alleviate nerve disorders like Angelman's disease and Prader-Willi syndrome.

Example: Yale New Haven Hospital became one of the first places in the United States to treat congenital heart defects with a stem cell-related therapy.

Example: Profs. Jeffery Kocsis and Stephen Waxman at Yale, along with their Japanese collaborators, have made significant progress in using stem cells in clinical trials to treat patients suffering from stroke and paralysis.

Example: Diane Krause of Yale has used pluripotent stem cells to fight leukemia.

Example: Haifan Lin's stem cell research has led to the discovery of a new way to block the growth of breast and other cancer cells without affecting normal cells.

But when I think of Connecticut stem cell research, I remember first and always Xiangzhong "Jerry" Yang, who helped found the University

Haifan Lin (http://www.cellbiology.yale.edu/people/haifan_lin.profile).

[2]http://regenerativemedicine.uchc.edu/faculty/bios/dealy.html

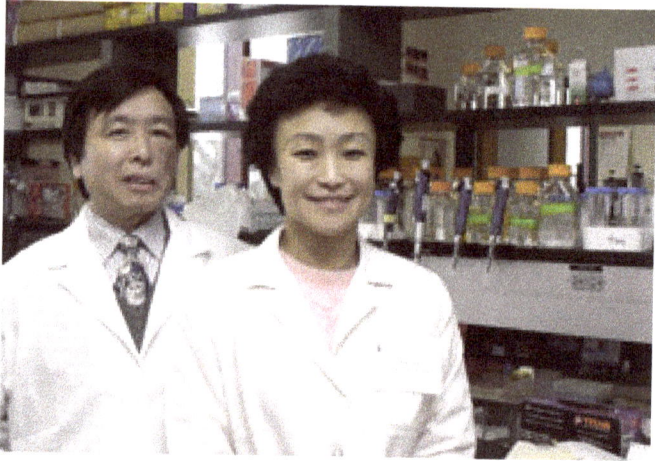

Jerry Yang and Cindy Tian (UCONN photo).

of Connecticut's Center for Regenerative Biology. He worked on "…producing tissue to be used in heart surgery, organ replacement, and repair of birth defects."

I knew Jerry only briefly; he was dying of salivary gland cancer. Operations had cut into his facial muscles and made speech difficult, but he remained cheerful and enthusiastic, talking a mile a minute. With every fiber of his being, he fought to improve the lives of others, and he continued working in his lab until he died.

Dr. Jerry Yang is gone from us now, but his wife Dr. Cindy Tian continues his work, sponsored by the Connecticut Regenerative Medicine Research Fund, as they turn theory into therapy, to save lives and ease suffering.[3]

A grant should be established in memory of Jerry Yang, and his unsurpassable dedication to research.

His life embodied the Connecticut commitment to cure.

---

[3] https://en.wikipedia.org/wiki/Xiangzhong_Yang

30 IN MEMORY OF BEAU

When Vice President Joe Biden stood beside the coffin of his son Beau, it seemed the whole world took a breath. Every parent knew what he must be feeling: one of the most powerful men on Earth, and yet he could not save his son.[1]

Beau Biden died of a malignant brain cancer called a glioblastoma: the malady which took the lives of Ted Kennedy, Gene Siskel, Susan Hayward, George Gershwin, many others, and which threatens the life of Senator John McCain.

Glioblastomas kill about 15,000 adults each year: a deadly condition. Even with the best treatment available — surgery, radiation, chemotherapy — survival averages only about 15 months.[2] Untreated? Less than that. From diagnosis to death may be as little as four months.

Surgery? "Glioblastomas are often difficult to reach, due to the complex ... pathways of the brain. Worse, these cancers do not have clean boundaries. The main tumor mass can be surgically removed, but even the best operations leave many invading cancer cells scattered throughout the brain, continuing their deadly growth." — Karen Aboody, MD, personal communication

Radiotherapy can have negative side effects. Chemotherapy drugs may be blocked from reaching the brain by the body's own defense, the

---

[1] http://www.sandiegouniontribune.com/news/health/sdut-beau-biden-brain-cancer-2015jun16-story.html
[2] http://www.abta.org/brain-tumor-information/types-of-tumors/glioblastoma.html?referrer=https://www.google.com/

Joe Biden (http://nypost.com).

blood-brain barrier. And when they do get through, such drugs may do harm to the healthy portion of the brain as well as the tumor.

At City of Hope (COH) National Medical Center in Los Angeles, Karen Aboody, MD, Professor of Neurosciences and Neurosurgery, and Jana Portnow, MD, Associate Professor of Oncology, are searching for a better way. Many at COH have contributed significantly to the advancement of brain cancer research: Larry Couture, PhD, Behnam Badie, MD,[3] Christine Brown, PhD, and more.

But Aboody and Portnow capture the imagination: women warriors working together for years with very different and complementary skills.

Dr. Aboody is a strategist; she studies the battlefield, learning the enemy's strengths and weakness, planning campaigns for its defeat.

Dr. Portnow is a battlefield commander: Principal Investigator of the neural stem cell clinical trials, she designed the protocol and oversees the running of the study.

---

[3] http://www.businesswire.com/news/home/20161228005353/en/City-Hope-Researchers-Achieve-Remission-CAR-T-Cell

Beau Biden (en.wikipedia.org).

Aboody's early work at Harvard led to a curious fact: one kind of stem cell, a neural stem cell (NSC), is attracted to malignant and invasive tumor cells. Even if injected into the brain at a distance from the tumors, NSCs will migrate through normal tissue to seek out glioma sites. Dr. Aboody came to City of Hope 13 years ago to study this, and try to advance it toward patient trials.

Remember the Rambo movies, where the hero fired exploding arrows into the enemies' ammunition? This is a little like that. The NSCs would shoot in toward the brain cancer. When they arrived at just the right place, their presence would trigger a previously-inactive drug and kill the surrounding cancer cells.

That anti-cancer drug, inactive until the NCSs arrive, could be swallowed as a pill.

If the procedure works, the stem cells would put localized chemotherapy exactly where it has to go: sparing normal tissues from toxicity and potentially decreasing the side effects so common to chemo.

How would it be tested?

Enter Dr. Portnow, leading the charge, working with the patients. The first-in-human safety study, completed in 2013, showed no negative responses from the body's immune system. It also provided proof that the anti-cancer agent was activated where it was supposed to: at the brain tumor site.

Karen Aboody (cityofhope.org).

In 2014, Dr. Portnow launched a Phase 1 dose escalation, multi-treatment clinical study, currently ongoing at City of Hope.

Here is how it works:

At the time of surgery, to remove a recurring brain cancer, the patient receives injections of the neural stem cells. The patient also has a small catheter inserted in the brain, for further rounds of outpatient treatment. After each stem cell infusion, the patient goes home and takes the inactive drug in pill form for 7 days. The stem cells convert it to an anti-cancer agent in the brain.

This current trial should be completed in two years.

And the next generation? Drs. Aboody, Portnow, and Couture have received an $18 million grant[4] from the California stem cell program to bring a second generation NSC treatment to patient trials. This treatment may be more potent than the first, and (importantly) can also be applied to cancers outside the brain. It was a five-year effort, leading to the filing of a new IND (Investigational New Drug) application to the Food and Drug Administration (FDA).

Three patients have received the new therapy; more are being recruited.

---

[4]https://www.cirm.ca.gov/our-progress/awards/stem-cell-mediated-therapy-high-grade-glioma-toward-phase-i-ii-clinical-trials

Jana Portnow (cityofhope.org).

Help is on the way, to defeat a previously incurable condition.

When the first patient was enrolled at the CIRM alpha site at City of Hope, Maria Millan, Senior Director of Medical Affairs at CIRM, had this to say:

"This is the start of something truly unique ... by working together, providing collective expertise, efficiencies and critical resources; we can help accelerate the development of stem cell treatments for patients with unmet medical needs."[5]

When vice President Biden grieves for his son, and I am sure he thinks of him almost every minute, I hope he will take comfort in this fight: as California seeks to end a terrible condition.

---

[5]https://www.cityofhope.org/news/first-patient-treated-at-new-cirm-funded-stem-cell-clinic

31 TO RELOCATE ALLIGATORS, OR TURN A COUNTRY ON TO BIOMED?

The first one is easy: I know, I have done it.

At Marine World we had to move a group of alligators. The plan was for a man with a catch-pole to loop a noose over each reptile's jaws and pull them shut. Once the top and bottom jaws come together, they seemed confused, and we could control them easily.

This worked very well, with the smaller ones.

But we saved the largest for the last, a ten-foot animal with a non-cooperative attitude. It was up on all fours, tooth-dripping jaws yanking back and forth, like aiming a gun. Its breath made steam in the chill morning air: "HAAAAHHH."

With its attention diverted, I tiptoed around the reptile, stepped over its back, and lunged down, pinning the snout to the floor.

The alligator froze. I worked my fingers under its jaws. The vet came over quick and gave me the world's most unnecessary advice.

"Hold on tight," he said.

Hold on tight? If I hung on any tighter I would have left fingerprints! I maintained position until the vet wrapped the cloth adhesive tape around a couple times.

Then we picked the gator up like luggage, and set it with the others on the truck.

It was not difficult — a gator has no strength opening its jaws; closing is the power.

(Note: this technique applies only to alligators; crocodiles have differing opinions.)

But to convert a country to biomed? For that I can only point to Singapore.

Philip Yeo (en.wikipedia.org).

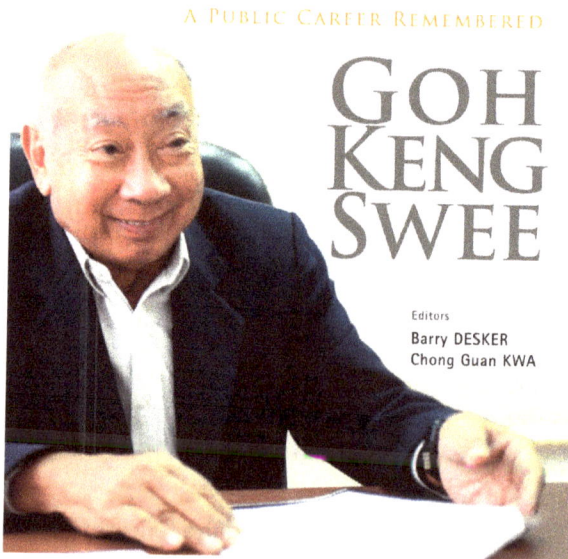

Goh Keng Swee ("A Public Career Remembered").

When Singapore became independent in 1965, it was a third world country with few advantages. It could have been exploited as a source of cheap labor, taken advantage of by other countries. Instead, it became a prosperous nation with a well-educated population, an economy based on science, and lots of jobs.

"In little more than a decade, Singapore established a thriving biomedical industry from scratch ... In 2000, it started to invest in a Biomedical Sciences initiative ... by 2012, this industry had grown to ... $24 billion (US) dollars ...."[1]

"Leading companies (Pfizer, Abbott, GlaxoSmithKline, Lonza, MSD, Novartis, and Sanofi-Aventis) have made Singapore their global manufacturing base ..."

One key policy was to seek information from outside experts, and adapt their ideas to fit Singapore.

For instance, in 1970, an Israeli scientist, Dr. Meir Ben Zvi, offered a detailed plan to develop Singapore's scientific personnel by sending students overseas to study. The advice was taken, and modified.

A civil servant, Philip Yeo, called that concept "guppies and whales," sending out "guppies," students to learn, and bringing in "whales," international experts.

Yeo worked closely with the visionary Finance Minister Goh Keng Swee, who gave him considerable leeway. Look anywhere in Singapore's research world, and you will see both men's fingerprints.

Yeo brought change, Goh Keng Swee backed him up. If anyone complained, the Finance Minister would say "What can I do? That's Philip!"

Example: Yeo changed the name of Singapore's biomed division, from something no one could remember, (the National Science and Technology Board, NSTB) to something they could not forget — A*STAR (Agency for Science, Technology and Research) — A*STAR is the top grade a Singapore student can earn.

A brilliant young scientist, Tsao Chieh, developed a life-threatening disease. Yeo arranged for therapy to be brought in, but Tsao had an allergic reaction, and died.

"I was watching a young life die before my eyes," said Philip Yeo, "And there was nothing I could do ..." But he did not forget. When Abraham Lincoln first saw American slaves, chained neck to neck, on their way to being sold, he was said to have promised, "If I ever get a chance to hit that, I'll hit it hard." And so it was with Philip Yeo; he remembered, and he waited, and he changed the world.

---

[1] http://www.worldscientific.com/worldscibooks/10.1142/9417

Edison Liu (Genome Institute of Singapore).

Lim Chuan Poh (a-star.edu.sg).

In 1983, Nobel laureate Dr. Sydney Brenner gave a hugely influential series of lectures about biomedicine. Philip Yeo and Goh Keng Swee arranged for those lectures, and worked to make their message become real. Working with Dr. Brenner, Yeo made a list of 100 top scientists and tried to bring them to Singapore.

The first top scientist he successfully contracted was a young Dr. Edison Liu, to be the founding executive director of the new Genome Institute ...

Remember the bird flu epidemic? Called SARS, the Severe Acute Respiratory Syndrome, "It was a deadly global epidemic ... controlling the spread of this highly communicable disease became paramount."

It was Dr. Liu's baptism by fire. He mobilized the scientists.

"I called on everyone to drop what they were doing to work on this problem ... In the morning of each day, we met with the team leaders and said, "Okay, what are you going to do today?" At the end of the day, we met with the team leaders again and said, "What happened today?" The gene sequencing team worked 24/7. I mean, they slept in the hallway."

Singapore developed a testing kit for SARS, and was "cited by the World Health Organization as an exemplary country in its response."

"That experience made it obvious to decision-makers that their investment in biomed was not just to get a cash return, it was (also) for national security ..."

Hepatitis B? Thanks to Singaporean efforts, that condition can now be diagnosed in two days instead of two weeks.

Another biomed champion is A*STAR Chairman Lim Chuan Poh, formerly a Lieutenant General in the Singapore military.

At the agency's 20th anniversary, he said:

"In the biomedical effort, we (built) capabilities for one big industry cluster. This was ... a new approach, changing the research landscape conceptually."

The visual reality of the "one big industry cluster" was a series of 13 incredible buildings, Biopolis and Fusionopolis, designed by one of the world's greatest architects, the late Iraqi-Brit designer Zaha Hadid.[2]

Two million square feet of shining steel, reaching to the sky, designed to both meet the needs of science and to be a living quarters; scientists could get up in the morning, have breakfast, kiss the family goodbye, and be at work in 30 minutes!

And how is their economy working, in terms of jobs?

Singapore's unemployment rate is less than 2%.

---

[2]2016-06-25-1466856083-9201983-biopolispic.jpg

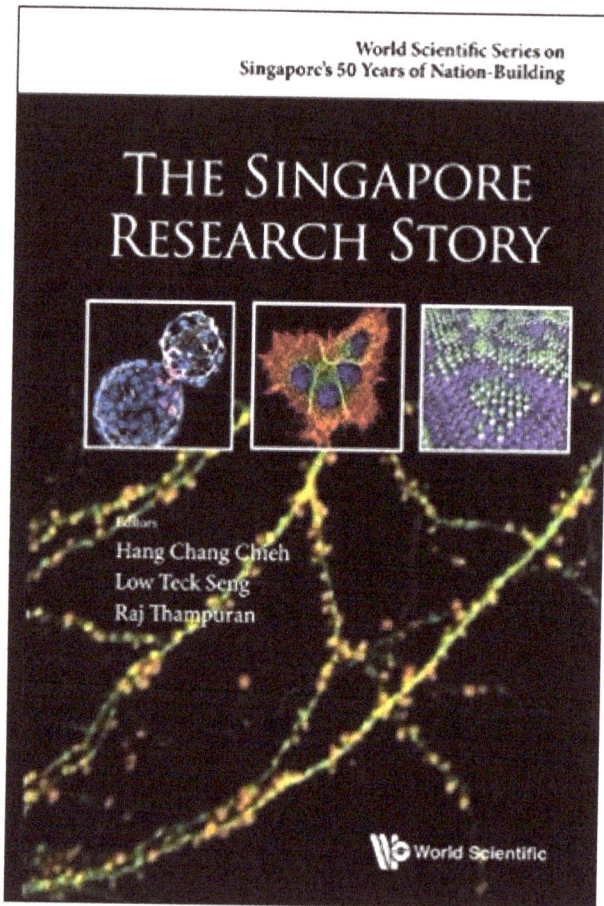

World Scientific Publishing.

The island nation's success story shows what can happen to a country, when they unite in support of science.[3]

Note: If you want to know more, there is a book I recommend highly, by the same publishers who did mine.

The Singapore Research Story, edited by Hang Chang Chieh, Low Teck Seng, and Raj Thampuran, World Scientific Publishing, Inc. 2016.

---

[3]http://www.bbc.com/news/av/business-24799516/singapores-biomedical-sector-growth

32 WHALE SHARKS AND OUTER SPACE

Swimming toward me were two of the biggest sharks in the world.[1]

Slow-moving, majestic, harmless, (and in any event on the other side of a thick plate glass window), these were whale sharks, each about 20 feet long. "Only teenagers, half-grown," said the diver through a tank microphone.

As a former professional diver, I had heard about these gentle giants all my life, and would have loved to spend a day at the magnificent Georgia Aquarium.

But I was on a mission.

I stood in awe for just a moment, posed for a souvenir photo that made me look like I was standing in the tank — and then rushed back —

To the Atlanta Hyatt Regency, and the World Stem Cell Summit.

Presented by Bernard Siegel of the Genetics Policy Institute, the Summit takes all year to build. It unites advocates, scientists, entrepreneurs and government officials from 35 nations, and this year from outer space too![2]

Did you know stem cell research will be done at the International Space Station? The Center for the Advancement of Science in Space (CASIS) was issuing a request for applications to do stem cell experiments in

---

[1] https://www.georgiaaquarium.org/experience/explore/programs-activities/animal-interactions/journey-with-gentle-giants/swim-with-whale-sharks
[2] http://worldstemcellsummit.com/about-summit/

Whale Shark at Georgia Aquarium (deanoinamerica.wordpress.com).

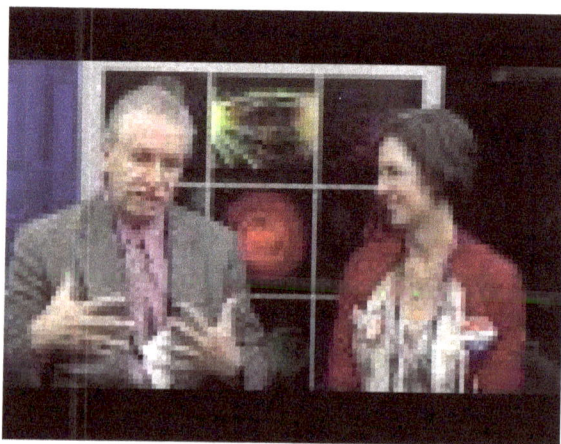

Author at World Stem Cell Summit: https://stemcellbattles.net/videos/

microgravity. How would stem cells react in outer space, with no gravity to squash them down?[3]

A more down-to-earth event involved the Food and Drug Administration.[4]

The President of the California stem cell program stepped to the microphone:

"There is an excessively long…pathway to get an Investigational New Drug (IND) approval from the FDA (just to get started, before testing a therapy). For non-cell therapies it takes 3–4 years to get an IND. For cell therapies? 6-8 years."

"We have had the current FDA regulatory structure in place for 15 years, and in all that time not one stem cell therapy has been completely approved. Not one …"

"We are not anti-regulation, not anti-FDA, not calling for the removal of rules and regulations for stem cell therapies … But right now we are being so careful about safety to ensure patients are not put at risk, while patients are actually dying …"

— Dr. Randy Mills, CIRM President.[5]

What if that delay could be lessened? Not the safety tests: those are crucial. But, what about removing one of the three "efficacy" (does it work?) tests? That would shave years off the process, and perhaps save lives.

FDA representatives Robert Cailiff and Celia Witten explained the size of the situation — FDA regulations affect 25% of the USA's gross domestic product![6]

[3]http://www.iss-casis.org/Home/tabid/110/ArticleID/98/ArtMID/625/Default.aspx

[4]http://blog.cirm.ca.gov/2015/12/16/doing-nothing-is-not-ok-a-call-for-change-at-the-fda/

[5]http://blog.cirm.ca.gov/2015/12/16/doing-nothing-is-not-ok-a-call-for-change-at-the-fda/

[6]http://www.fda.gov/downloads/ICECI/EnforcementActions/EnforcementStory/UCM129822.pdf

Alain Vertes (researchgate.com).

And the budget does not meet the need; FDA problems increase, but not their manpower to deal with them.[7] Their goal was ours: "... safe and effective therapies that can be delivered reliably."

America needs to be sure the FDA has the funds to do its job.

Listen to the (free!) speaker videos.[8]

The International component was always absorbing.

Soft-spoken Alain Vertes of Switzerland spoke of the importance of government-sponsored "incubator programs" for start-up research: Vertes is author and editor of the new book "Stem Cells in Regenerative Medicine: Science, Regulation and Business Strategies." I purchased a copy: excellent.

Israel was "a nation has which systematically fostered stem cell research," said Avi Treves, deputy director of the Sheba Cancer Research Center.

---

[7] http://www.nytimes.com/2015/04/08/us/food-safety-laws-funding-is-far-below-estimated-requirement.html?_r=0

[8] http://worldstemcellsummit.com/video-resources/

How about 40 Maple Leaf companies working together? Canada's Michael May shared the vision of the Centre for Commercialization of Regenerative Medicine (CCRM), a 40-company consortium "bridging the gap between academia and industry ... regenerative medicine commercialization."[9]

China had set up a special fund, so that 450 million renminbi (roughly $64 million USD) a year would go into stem cell research ...[10]

Amazing moment: I was heading for the escalator (the Summit is spread over several floors) when I bumped into Rich Lajara, a wheelchair warrior I met recently. I had known about him for quite some time, he being the first person in California to receive embryonic stem cells in the Geron safety trials.

We spoke for a bit, and then, to my amazement, Rich did a wheelie on his wheelchair, and charged full speed at the moving escalator. Front wheels in the air, he rode the escalator all the way up to the top ...

He wrote me a note after the Summit.

"You had asked me what my take away was from the Summit. I'm not a victim of an unfortunate situation; I'm a survivor with unimaginable opportunities. I'm grateful when the stars aligned I was in the wrong spot at the right time. There is not a single person or family that has not been or will soon be afflicted with a condition regenerative medicine may help." — Rich Lajara, personal communication

The advocates were what I noticed most this year: people like Davis George.

To understand the power of the young Mr. George, you have to meet his brother. Jacob George was terribly injured by a fall off a tractor when young. He had been paralyzed almost to the point of death. His body is near motionless, to the point where he can barely open his eyes.

But his face retains some muscle control. After you are with him a while you can see the expansion of his cheeks and mouth, and you recognize

[9]http://www.nce-rce.gc.ca/NetworksCentres-CentresReseaux/CECR-CECR/CCRM_eng.asp
[10]http://knowledge.ckgsb.edu.cn/2014/05/05/technology/stem-cell-research-in-china-regenerative-economics/

Rich Lajara (blog.cirm.ca.gov).

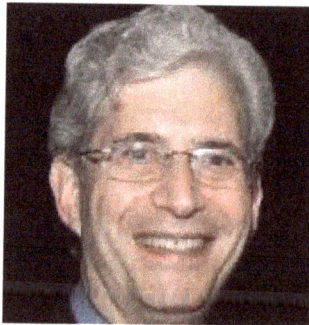

Bernard Siegel (linked-in.com).

happiness. When I yelled out to him from the speaker's platform, "Hey, Jacob!" he smiled.

Later, Jacob's little brother was pushing him in the wheelchair, rapidly. I asked him did his brother like going fast, and the answer was yes, he had once owned a Spider Lamborghini, and sometimes he imagined himself back in the Spider …

And Davis George? Slender, and extraordinarily young-looking, he was the most promising patient advocate I have seen in years. A debating champion at Duke University, he is clear-spoken, dedicated, and if anybody uses politics to get in the way of his brother's hope of cure, well … they're going to meet Davis George!

33 MR. SCIENCE GOES TO WASHINGTON?

On June 15, 2017, Hans Keirstead, PhD., announced his candidacy for Congress, U.S. Representative, District 48, Costa Mesa, California.

Should we send a scientist to Washington?

Absolutely. We need to balance the science non-believers, who might not recognize global temperature change if an iceberg crushed their beach house!

Keirstead is not only a world-class scientist, but a man firmly rooted in the day-to-day problems of real people. For years, he and his father have quietly arranged for literally tons of vitamins and medicines, and thousands of wheelchairs, to be donated to some of the poorest people in Africa.

An extraordinary individual, Hans has a third-degree blackbelt in Tae Quan Do, is a helicopter pilot, and has brought the world closer to cures for cancer and other chronic diseases.

I have known him for nearly twenty years. As you know, when Roman was paralyzed in a college football accident, we passed a medical research law in California, the Roman Reed Spinal Cord Injury Research Act of 1999.

Keirstead was one of the first scientists funded by "Roman's Law." He used stem cells to "re-insulate" the damaged nerves in a paralyzed rat. Her name was Fighter, and at first it made you cringe to see the white rodent trying to walk, dragging her useless hind-limbs behind her. But after she got the stem cells, she scampered across the purple plastic swimming pool (her play area) with her tail held high. Motion had been restored.

Hans Keirstead (UCI.edu).

Today, in FDA-approved clinical trials run by Asterias Biotherapeutics, that same therapy is benefiting human beings. As mentioned previously, in the last cohort of six paralyzed people to receive the experimental stem cells, all recovered hand control. For a paralyzed person to regain the use of the hands? That is a dream come true, and it would not have happened without Dr. Hans Keirstead.

In addition to paralysis, Hans is also challenging late stage cancers, immune disorders, motor neuron diseases, and retinal disorders.

Where he is, accomplishments follow.

When he received his PhD. from the University of British Columbia, Canada, it won the Cameron Award for the outstanding PhD. thesis in the country.

At the University of Cambridge, Hans was elected Senate Member, the youngest member ever to receive that honor.

As CEO of Aivita Biomedicine, Hans has been trying to raise money to fight cancer. So, being Hans, he came up with a new way to fund science: a face cream that appears to reduce wrinkles. He has pledged that every dollar of profit from the anti-wrinkle cream will go to the fight against ovarian cancer.

When California was trying to decide whether or not to support stem cell research, Hans went on the 60 MINUTES TV show, showing how paralyzed rats could walk again. He served on the Science Advisory

Board for Proposition 71, Bob Klein's citizens' initiative, which led to the California stem cell program. And when Nancy Reagan wanted to know about stem cell research, she reached out to Hans.

It is difficult not to like Hans. As someone who genuinely enjoys and respects the accomplishments of others, Keirstead inspires friendships that last. The people working with him now? Often the ones he began with. Talk to Gabriel Nistor, who has been with Hans 17 years.

For instance, I am a behind-the-scenes patient advocate. I have been at it so long I have been called the Grandfather of stem cell research advocacy.

But still I am a backup man, a face in the crowd, someone you are not likely to remember. My son? That's different. When Roman is "on," he practically sheds light — him you can never forget!

The other day at a stem cell conference Roman developed, I was reading a book to my 8-year-old grand-daughter Katherine, who had been so patient with all the talk.

And Hans approached and said to her:

"Did you know your Grampa changed the world?"

That is Hans Keirstead: a positive force, he appreciates and brings out the best in people. In Washington, he wants to work cooperatively, to solve the world's most difficult problems.

Is that not the kind of person we want in our nation's capitol?

34 WHEN OKLAHOMA IS NOT OK

My cousin Michael Gallegos lay in a hospital bed. Under the covers, a tube ran from his right foot down to a heavy plastic bag of dark liquid.

Mike has diabetes, and the circulation in his right foot had been shutting down. The skin was turning black. So the doctor had taken a felt marker pen and drawn a circle around the base of the toe.

"If the darkness gets to this line," he said, "We will have no choice but to amputate."

Today the toe was gone.

"Take a look," said Mike, pulling the covers back.

A two-second glimpse, and I turned my head away.

How should we reward a stem cell researcher who found a cure for diabetes?

In Oklahoma, the compensation might be prison: "… incarceration in the county jail for a term not to exceed one year."[1]

A bill called the "Protection of Human Life Act of 2015," HB 1379 by Representative Dan Fisher, recently passed the Oklahoma House of Representatives. This bill "would make it a … crime to perform embryonic stem cell research in Oklahoma."[2]

The author of the bill, Republican Dan Fisher calls embryonic stem cell research "murder." Listen to what he said last year, about a similar bill he wrote, HB 2070:

"As Baptists we uphold the value and sanctity of life," said Fisher, Pastor of Trinity Baptist Church in Yukon. "We believe that life should

---

[1] http://www.ecapitol.net/viewtext.wcs?HB1379_CS~55th

[2] http://newsok.com/article/5390594

be protected from the moment of conception (fertilized egg — dr) ... anything that stops that (process) ... is called murder."[3]

Mr. Fisher is entitled to his religious beliefs. But to impose them on all of us?

What precisely is the "life" his bill wants to protect?

"No person shall ... conduct ... research that destroys a human embryo ..."

Webster's Dictionary defines an embryo as: "especially: the developing human individual from the time of implantation to the end of the eighth week after conception."[4] (emphasis added).

Representative Fisher has his own definition, a political one; according to him:

"Human embryo" means a living organism of the species homo sapiens at the earliest stages of development, including the single-celled stage, that is not located in the body of a female" (emphasis added). So no implantation ...

Of this one cell embryo, outside the woman, Representative Fisher said:

"The only thing that this embryo needs to become a fully developed baby is time and nourishment."[5]

The "only thing ... the embryo needs?" What about the womb? Does not Mr. Fisher consider the uterus a requirement? Without its nurturing shelter, it is biologically impossible to make a child.

So are the rights of a single cell more important than millions of people with disease and disability, who might be helped by a stem cell therapy?

Let us be clear where embryonic stem cells come from: fertilized eggs left over from the In Vitro Fertilization (IVF) process.

A couple contributes their biological materials, and the doctor puts together 15-20 fertilized eggs. Of these, the strongest one or two will

---

[3] http://www.rawstory.com/rs/2012/04/religious-groups-push-to-revive-oklahoma-fetal-personhood-bill/

[4] http://www.merriam-webster.com/dictionary/embryo

[5] http://www.okcfox.com/story/24822815/ok-house-votes-to-ban-embryonic-stem-cell-research

be put into the woman's womb, in hopes it will implant in the uterine wall, and become a child.

But what about the others? These microscopic bits of tissue, so small they could fit comfortably on the point of a needle, may be discarded, or frozen and stored, or given to another couple or donated to research.

How badly is cure research needed?

In the year 2012, 29 million Americans had diabetes, roughly 9% of the population.[6]

Estimates of the number of paralyzed folks run as high as 5.6 million.[7]

And 11 million Americans are losing their sight from macular degeneration (age-related progressive blindness).[8]

Why did I choose these three groups of people?

Because an embryonic stem cell therapy for each of those conditions — diabetes, paralysis, age-related blindness — is in human trials right now.

1. Diabetes: ViaCyte Inc. has a sturdy "teabag" in which the precursor stem cells release insulin. This is being put under the skin of patients.[9]
2. Paralysis: Asterias Biotherapeutics is enrolling newly paralyzed patients to receive embryonic cell-derived precursors which will turn into oligodendrocytes, to re-insulate damaged nerves in the spine, permitting recovery of function.[10]
3. Age-related blindness (macular degeneration): Advanced Cell Technology ran the first series of tests (minor doses for safety alone) and found that ten of 18 patients developed improved vision with the embryonic-derived stem cells.[11]

---

[6]http://www.diabetes.org/diabetes-basics/statistics/
[7]http://www.webmd.com/brain/news/20090421/report-nearly-five-point-six-million-americans-paralyzed
[8]http://www.brightfocus.org/macular/about/understanding/facts.html
[9]http://viacyte.com/clinical/clinical-trials/
[10]http://www.prnewswire.com/news-releases/asterias-biotherapeutics-initiates-patient-enrollment-for-phase-12a-clinical-trial-of-ast-opc1-in-newly-injured-people-with-complete-cervical-spinal-cord-injury-300045731.html
[11]http://www.thelancet.com/journals/lancet/article/PIIS0140-6736(14)61376-3/abstract

Doug Cox (Integris Health).

Sadly, HB 1379 passed the Oklahoma House with a Republican majority (80-13).

One Republican, Doug Cox, was the voice of conscience:

"By passing HB1379 which criminalizes embryonic stem cell research, the Oklahoma House is sending the message to the scientific community that we would rather incinerate unwanted, unused embryos, rather than harvest a cell that could be used in research for a cure of diseases that affect mankind." — Representative Doug Cox (R), personal communication.

And what about the biomedical aspects of a ban on research?

I reached out to the job-creating hub of the state, Greater Oklahoma City's Chamber of Commerce, and spoke to Mark Vanlandingham, vice President of Governmental relations.

I expected Mr. V to talk numbers and economics and jobs.

But his first response was: "Is there no higher use for unwanted embryos than to just throw them away? Should we not use them to try and save lives?"

He had the numbers too: Greater Oklahoma City had 27,800 workers in the biomed field. The industry had generated six percent of the state's income, $4.1 billion last year. These researchers and technicians are Oklahomans, many with families.

Oklahoma State University is a world-class institute, and it seems a shame to deny students the right to research medical breakthroughs.

Will Oklahoma become infamous as the capital of anti-science? I hope not.

The great Rogers and Hammerstein musical OKLAHOMA contains the phrase, "Oklahoma — ok!"

Let us hope that phrase will one day symbolize the state's new attitude toward science in general, and stem cell research in particular.

FLASH: HB 1379 just "died in (Appropriations) committee …"[12]

---

[12] https://legiscan.com/OK/bill/HB1379/2016

35 JAMES BOND AND MELANOMA

In the classic movie GOLDFINGER, James Bond is tied on his back to a table, legs apart, while a crackling buzzing laser beam is about to saw him in half. Bond struggles helplessly, then calls to the villain: "Do you expect me to talk?

"No, Mr. Bond," says Auric Goldfinger, mildly surprised: "I expect you to die."

Naturally JB talks his way out of the situation. Movie villains love to brag and show off before the one man on Earth who can spoil their plans. Bond is released, the world is saved, and the franchise continues.

But in that instant, when the villain says: "No, Mr. Bond, I expect you to die," there was a distinct chill, as if Death had entered the room.

Once, cancer diagnosis was a threat from which there was no escape. But today, with early action and encouraging research, we do have a chance.

Quick quiz: do you know the ABCDE's of skin cancer?

"**A**symmetry — one half different than the other; **B**order irregularities — the edges are ragged; **C**olor differences — varying shades of tan, brown or black; **D**iameter — bigger than a pencil eraser; **E**volving — the mole noticeably changes."[1]

What should you do, if you get a mole on your skin and it has any of the above?

Take it to the doctor. Caught early, skin cancer can be handled: no big deal.

---

[1] http://www.skincancer.org/skin-cancer-information/melanoma/melanoma-warning-signs-and-images/do-you-know-your-abcdes

Antoni Ribas (newsroom.ucla.edu).

The doctor might look at the mole on your cheek and say: "Take this off?" You say, "Yup," he numbs the skin, half an hour later, it's gone. You wear a bandage for a couple days, show off the scar to your friends, that's it.

But if you wait too long?

A deadly skin cancer called melanoma can metastasize and spread: traveling through the blood or lymph system, invading lungs, liver, intestines, brain, or eyes.

"World-wide, more than 160,000 new cases of the skin cancer melanoma are diagnosed each year. The vast majority are caught early enough that they can be cured by surgery. But when this tumor spreads … it becomes highly resistant …"[2]

At that point, even surgery, chemo, radiation and hormones may not help; death may come in months.

We need a way to fight.

Enter Antoni Ribas, M.D., Ph.D.

His father was a doctor, his grandfather was a doctor, and his great-grandfather was a doctor, so he was going to be a doctor, too, right?

Well, no. He had signed up for engineering college, because he wanted to make things. But there was too much math involved, and he did not like math!

---

[2]https://www.cirm.ca.gov/our-progress/melanoma-fact-sheet

Because he did not like math, the world gained an important anti-cancer fighter and an organizer, too. The Spanish-born scientist is the "director of the Parker Institute for Cancer Immunotherapy Center at UCLA, which brings together the nation's leading cancer centers to develop new therapies for (fighting cancer)."[3]

And what is the Ribas approach to battling melanoma?

He manipulates blood cells to "continuously generate melanoma-targeted immune killer cells, hopefully providing protection against the cancer ..."[4]

CIRM is helping him to the tune of $19 million dollars ($19,875,776). Naturally the money comes with conditions, mutually-selected milestones; if the treatment does not work, the money stops.

But Antoni Ribas will have his chance; California is backing him up.

Cancer should be an inconvenience, not a death sentence.

---

[3] http://www.onclive.com/publications/oncology-live/2016/vol-17-no-13/ribas-builds-foundation-for-immunotherapy-success?p=2
[4] https://www.cirm.ca.gov/our-progress/awards/genetic-re-programming-stem-cells-fight-cancer-0

36 NEUROLOGICAL DISEASES VS. CALIFORNIA

First, a seeming digression.

As previously mentioned, in my youth I worked as an aquarium diver for Marine World Africa USA, in Redwood City, California. Five days a week I would swim in the tanks full of wildlife, creatures of a man-made sea.

The most beautiful exhibit was a six hundred fifty-thousand gallon tropical reef display. It had giant groupers big as Volkswagens, and tiny cleaner fish which swam in and out of their mouths: plus the Technicolor beauty of angelfish, surgeonfish, damselfish, wrasses, silver tarpons, alligator gars and many more.

And then one day the fish began to die. I did not know why; neither did the vet. I just carried them out of the tank, one by one, these fish I knew as individuals.

At first the turtles ate the dead fish, and this was somehow comforting, for at least someone got benefit. But soon there were too many for even them.

At last the groupers came out of their cave, lay head to head in the sunlight for several hours as though saying goodbye. Then they swam to different corners of the tank and slowly lost their color: we dragged their giant bodies out with ropes.

After three weeks, the tank was barren, empty, like the landscape of the moon. We turned off the heaters, and changed to a cold sea collection of local fish.

We never found out what had killed that underwater neighborhood.

But what if there had been one solution, to save the lives of many?

Hold that thought.

Author and orcas (personal).

Now. Consider nerve diseases like Parkinson's, Huntington's, Alzheimer's and ALS (Lou Gehrig's disease): all incurable.

What if they had a common weakness: might a single medication defeat them all?

If you go to the California stem cell agency web page, www.cirm. ca.gov, and look up Steven Finkbeiner's project (he works at Gladstone Institute), you will find an amazing possibility.

First, what are we up against?

"A major medical problem ... is the growing population of individuals with (nerve) diseases, including Parkinson's (PD) and Huntington's (HD) ... These diseases affect millions of people, sometimes during the prime of their lives, and lead to total incapacitation and ... death."[1]

Millions of people incurably ill, with no expectations but death?

But in the very next sentence of his grant proposal:

"We propose to conduct ... studies to understand the common disease mechanisms of (these) disorders ... to develop effective treatments for these diseases ..."

---

[1] https://www.cirm.ca.gov/our-progress/people/steven-finkbeiner

I called up Dr. Finkbeiner, and found that the idea of hunting for similarities was a deeply held conviction. He called it "Identifying common threads," and suggested a stem cell technique to find them.

He would take skin cells from Parkinson's and Huntington's patients, and make a cellular model of the diseased cell. Then he would try a whole bunch of drugs on them, see if there was one which affected both diseases.

If a drug was found that was already approved by the FDA, they could skip the long years of testing and development — saving hundreds of millions of dollars, and make the therapy cheaper.

"The goal of our study is to identify common mechanisms that cause the degeneration of neurons and lead to most neurodegenerative disorders … we have made good progress in both developing … novel compounds and identifying potential genetic targets that could lead to … therapeutic strategies for patients with Huntington's and Parkinson's Disease."

One medication to defeat both Huntington's and Parkinson's and maybe other conditions like ALS, Alzheimer's and more?

What an incredible gift to the world that would be!

Meanwhile, back to the fish tank.

Sevengill shark (Dave Denardo photo).

Steven Finkbeiner (gladstone.org).

Having witnessed the horrors of a mass fish kill once, I never wanted to see it again, and became worried years later, when I saw a kelp bass lie on its side on the bottom of the tank, banging its head against the floor.

I knew this particular fish. It had a growth in its throat. When I tried to feed it, it flexed its gill covers to suck the food in, but the blockage would not admit the piece of cut-up herring.

And then, the ocean's answer to fish disease swirled in.

The brown and black-spotted sevengill shark circled closer and closer.

As if by accident, the rough pectoral fin touched the bass.

When it did not dart away, as would a healthy fish, the predator scooped it up, and tapped its teeth on the fish's back. Anything? But the protective spines did not stand up. The kelp bass seemed to relax. One violent shake of the shark's head, the fish was in half, and had become food.

In Nature's elegant simplicity, the disease (if such it was) was not allowed to spread. And the shark did not catch the illness.

The mass fish kill of years before did not repeat itself.

Will there one day be a drug or therapy as effective against nerve disease as the shark is against oceanic disease?

Steven Finkbeiner wants to find out.

37 DRIVING TO THE STORM

When Roman told me, he was going on a little trip, I said, "Oh, that's nice!" and went on with my chores. I figured he meant a couple-hour jaunt from Fremont to Sacramento, something like that, no big deal.

But his mother Gloria is more suspicious than I am, and managed to wheedle out of our paralyzed son that the "little trip" involved California, Texas, Alabama, and Louisiana, and that he would be driving all the way.

Complicating matters was a massive storm heading in, perhaps the most powerful ever recorded in this hemisphere ...

"That's why I have to go right now," he said, with perfect Roman logic.

Some might ask, why would I allow my paralyzed son to take such an incredible risk? Well, he is 40 years old. The days I can tell him what to do are long gone. Plus (if you know Roman), you know what it means when he makes up his mind.

Fortunately, brother-in-law Marty decided to make a two-man road trip out of it.

Even so, Roman's van was old, his powerchair was ancient, and Marty not only has diabetes but is also a stroke survivor, and cannot share the driving of the van.

The purpose of the trip? The same thing that has motivated Roman for the past 21 years to find, fund, and implement a cure for paralysis.

The research law named after him, the Roman Reed Spinal Cord Injury Research Act of 1999, has developed some amazing work. In fact, the first embryonic stem cell therapies ever done (now in clinical trials supported by the California stem cell agency) sprang from research initially funded by "Roman's Law."[1]

Want to electronically meet Kris Boesen, one of the people who took the risk and benefited? Check out his video.[2]

But still my son is paralyzed; and still he seeks the cure.

Roman personally owns rights to some amazing biomed patents, which he was seeking funds to develop; that was the purpose of the trip.

They took off, leaving his mother and I hip-deep in worry.

The first we heard was 48 hours later, a cryptic message: "Van broke down in Texas, phone giving trouble," and a picture taken through the van's front window. Framed by the steering wheel, the road ahead looked all cactus and sand.

We called him back, or rather we tried to, without result.

What to do? Drive down to Texas? Where exactly was he? Could we reconstruct his route? Should we call the police, or the Texas Rangers?

Six hours later, another email: "Marty fixed van. On our way."

The giant hurricane, perhaps the worst in history, howled in toward Texas.

Silence for two days.

Then, pictures of food in the email: crab cakes, jambalaya, bread pudding — "Louisiana food, unbelievable!"

Then: "Heading home. Lots rain." Heading home? From where?

And then at last, impossibly, Roman was in his living room, his children around him. Jay was talking soccer, while Katherine sat in her Dad's lap, explaining something complicated about school. Roman Jr., was too busy at Cal Berkeley, where he plays baseball and studies and eats. He is a 6'3" wall of rock nowadays — almost as tall as his Dad.

---

[1] https://www.cirm.ca.gov/our-progress/disease-information/spinal-cord-injury-fact-sheet

[2] https://americansforcures.org/2017/03/28/video-meet-kris-boesen-previously-paralyzed/

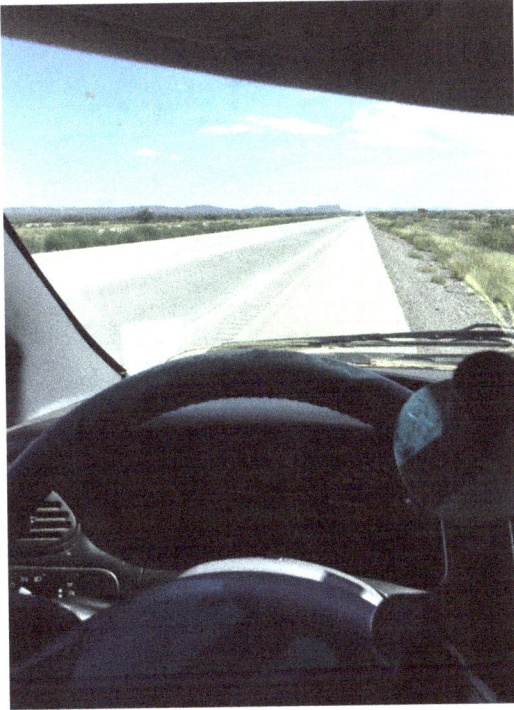

Cactus and sand... (Roman Reed photo).

Marty told how the van broke down.

"It made a noise like it could be the transmission," he said. "So we would just let it rest half an hour, drive ten minutes till the noise got bad again, then stop and cool down, did that for hours, until we came to a place that sold transmission fluid."

The weather was wild. Black fog slowed them to one mile an hour. An incredible lightning storm had surrounded them, hundreds of blasts of electricity.

But the hurricane did not catch up, though it dropped two feet of rain on states they had just left, flooding many roads.

The last 27 hours they had driven straight through. The inside of Roman's right wrist was black and blue from operating the hand controls

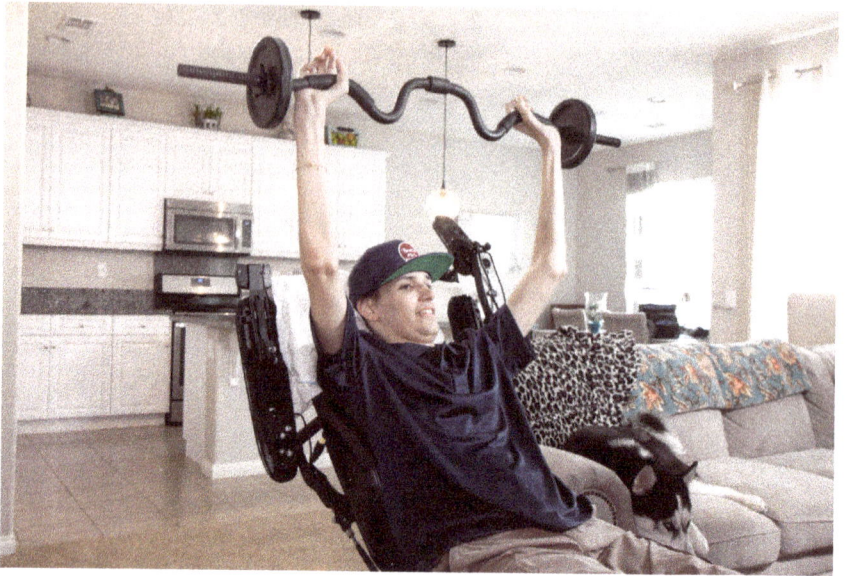

Kris Boisen (news.usc.edu).

Roman with sons Jason and Roman Jr. (personal photo).

so long. (On the steering wheel is a three-pronged device into which he inserts his wrist, to steer.)

Was the trip a success? The word was spread, Roman told me later.

"Daddy's sleeping", said Katherine.

Sleep tight, my son, and dream the dream of walking again.

38 DOOR INTO TOMORROW

"Optimism is the faith that leads to achievements. Nothing can be done without hope and confidence." — Helen Keller

On the 1,200 acres of the City of Hope (COH), life and death struggles are fought every day: against cancer, leukemia, blindness, diabetes, AIDS — battles which are not always won. Death can happen here. And yet, COH is not grim and gray.

Fountains and gardens rejuvenate the soul.

One tree was covered with fluttering colored paper "leaves," and every leaf was a note: a wish written by patient, friend or family.

"That's the Wishing Tree," said a cheerful nurse.

We were attending a meeting of the CIRM Alpha Stem Cell Clinics Network, in Duarte, California.

Sharing the adventure were Mary Bass and Yimi Villa, co-workers at Americans for Cures Foundation. Mary was on the phone constantly, knowing half the stem cell people in the world, and working on the second half. Yimi is a cheerful organizer who made sure everyone got where they needed to be. His scientific background included a stint at the University of California at Merced, where he received one of the CIRM Bridges grants for students.

Everything seemed touched with magic: the Los Angeles sky was bright and clean, after rain. Even the drive from the hotel was special. Beside me was vision expert Dr. Henry Klassen, who has vividly blue eyes. I asked him how he became a scientist. He said what clinched it for him was ... a pair of sunglasses?

Klassen's Dad was a biology teacher, who often took his young son to work.

"I liked it there, and picked up information almost by osmosis. So by the time I left the 6th grade I had a high school level understanding of biology. But I had headaches, and trouble with bright lights ... an opthmalogist told me I might have photophobia. When I left his office I had a prescription for sunglasses, which I thought was very cool, and an interest in opthmalogy."

Gathered in the Cooper Auditorium was a collection of experts.

Steve Rosen, Provost of COF, was the very first doctor to endorse Prop 71 and the author of more than 400 scientific papers. He was fiercely proud of COH, the largest bone marrow transplant facility in the world, providing more than 730 transplants in a year.

And over there was our friend Joe Gold! His 30 years in science included the glory days at Geron, when Dr. Tom Okarma had led the world's first development of an embryonic stem cell therapy, fighting paralysis.

Roman and I had attended the sad farewell dinner when Geron laid off its entire stem cell staff, including Joe. Their new management had quit stem cells for financial reasons. What a mistake that was! Today, the research Roman and Geron helped begin is moving forward at Asterias Biotherapeutics. And Joe Gold? He was director of manufacturing at COH's Center for Biomedicine and Genetics.

"The very latest," he said, "is that six out of six patients in the paralysis trials have regained arm function and hand control."

Imagine your fingers were paralyzed, numb, almost useless and then to have their function return courtesy of the California stem cell program.

Applause greeted the day's moderator, Patt Morrison. The six time Emmy Award winner was easy to find under a pink hat. But she was not there just for showbiz flair; she knew the issues; her contributions were sharp and to the point.

Randy Mills, CIRM's President and CEO, spoke of ambitious goals — aiming for 50 (fifty!) new clinical trials in the next year.

As the former President/CEO of Osiris Therapeutics, Dr. Mills had led the development of the stem cell drug, Prochymal, which treats graft-versus-host disease in children.[1]

---

[1] http://www.medicalnewstoday.com/articles/245704.php

The problem of <u>prices</u> was raised. As success mounts for stem cells, much has been made (and rightly so) about the predicted high expenses of therapies. Medicine is useless if you cannot afford it. Suppose, to cure blindness, it costs one million dollars per eye? Only the rich could afford to be cured.

Hopefully, the prices will come down as efficiencies occur. As Dr. Mills put it:

"The first widely available radio cost $450. And in 1977, the first saleable computer cost $1,300, a considerable amount in those days!"

Before we can have an affordable cure, we must first have a cure, or nobody will benefit. The trials were part of the answer.

"If we increase the number, quality and speed of the trials, we will get benefits to the patients more quickly," Mills said.[2]

High costs are not only for stem cells. The price of all health care is too high.

As Dr. Jennifer Malin of United Health Care put it:

"Most people spend about 10% of their income on food. But the total healthcare cost for a family is more than half their median income, and may equal that amount by 2018."[3]

Healthcare costs equalling a family's total income? How would they pay for everything else? We have to find ways to bring health care costs down!

CIRM's Geoff Lomax spoke briefly (too briefly) about the Alpha stem cells clinical trials network. As the point man for both the Alpha Network and the Stem Cell Center, he knew CIRM's accomplishments: 42 clinical trials under way.

And, he reminded us, funding had been approved for two more branches of the Alpha network.

Oddly, the announcement brought an ache of ... loss? Fear? Apprehension?

---

[2]https://www.cirm.ca.gov/patients/alpha-clinics-network/alpha-clinics-trials
[3]http://ascopubs.org/doi/pdf/10.1200/jop.2013.001017

Geoff Lomax (twitter.com).

There would come a day when the funding was gone.

What would we do without CIRM?

Unfortunately, I had little hope for help from the new folks in Washington. President Donald Trump's proposed first budget showed an astonishing cut in medical research: 18% less funding for the National Institutes of Health. The NIH was barely getting by as it was — flatlined at the same amount for years — and now he wanted it to be cut by almost one-fifth?

How does America feel about this? According to a major new poll, Americans oppose cuts in medical research by a gigantic proportion: 87-9%.[4]

87%? Almost nine of ten Americans did not want cuts in medical research. The California stem cell program embodies the will of our nation. It must be protected.

The conference was fast-paced; my attention was quickly yanked back.

Kristin Macdonald had intended to become a movie actress. She was beautiful (that has not changed), had an agent and was doing the work to make it happen. But she kept tripping and falling, once on the very

---

[4]https://poll.qu.edu/national/release-detail?ReleaseID=2444

staircase where the Academy Awards are presented. She broke her arm, twice. Her night sight went; colors faded. Her vision became like seeing through a slowly closing straw ...

"Retinitis pigmentosa," said the doctor. Kristin started using a white cane. She remembered the first time she heard someone say, "Look out, here comes a blind person!" and realizing they were talking about her ...[5]

But on June 20th, 2016, she was the first person in North America to have progenitor cells put in her eye. She had joined Dr. Klassen's clinical trial.

It was only a safety trial; the amount of transplanted material allowed by the FDA for this early effort was small: no benefits expected. However ...

"More light!" she said, describing the results. Even the awareness of light was an improvement, allowing her to recognize obstacles in her path.

Today she was an Ambassador for Americans for Cures, driving the fight in ways only an advocate can, spreading the word, raising awareness. As she puts it: "My eyesight is bad, but my vision is perfect!"

Other patient advocates spoke, fighters for cure like Pat Furlong, founder of Parent Project: Muscular Dystrophy.[6] Like many in the advocate community, she had very personal reasons to fight.

At the ages of 15 and 17, after years of gradual paralysis, her two sons died of Duchennes disease, a variety of Muscular Dystrophy.

"So I borrowed $100,000 and set out to defeat Duchennes," she said, naming the particular variety of muscular dystrophy. She has not succeeded — yet.

But today she is regarded as a world expert on Duchennes, and 46 other groups have sprung up to fight beside Pat Furlong.

---

[5] http://www.medicalrevolutionfilm.com/
[6] http://www.parentprojectmd.org/site/PageServer?pagename=About_media_presidentsbio

Kristin MacDonald (blog.cirm.ca.gov).

Pat Furlong (blog.cirm.ca.gov).

Cory Kozlovich bubbled with enthusiasm for using computers to gather data in a clinical trial, so that patient and doctor could know exactly what was going on.

An exciting new documentary was previewed: "Medical Revolution in Progress."

Christine Brown, Associate Research Professor at COH, gave the first understandable definition of T cells I ever heard, calling them: "White blood cells, soldiers of the immune system."

The battle against brain tumors was highlighted. Glioblastomas are a seemingly unbeatable foe; but here we saw proof they could at least be fought.

Benham Badie spoke about his father's death by brain tumor, and how that drove him on. After performing literally thousands of brain surgeries to remove tumors, he had joined COH and now, as the bio on the brochure said, he was using delivering "cancer-fighting drugs directly into tumors." We saw videos of brain tumors melting away, but still, new ones grew. The battle is not won.

So many top-flight people there: like John Zaia, leading the effort for a stem cell/gene therapy attack against AIDS.

There were founders of the biomed industry, like Louis Breton, co-founder (with Nobel Prize winner David Baltimore) of Calimmune and CellzDirect; and James Breitmeyer, President/CEO of Oncternal Therapeutics.

Jonathan Thomas, Chair of CIRM's Board of Directors, had not only been a Yale-educated lawyer, and public official, but also worked 15 years for the Crippled Children's Society of Southern California, now called AbilityFirst.

And today Bob Klein would be speaking on: "The Public's Role in Advancing Stem Cell-based Therapies."

Would today be the day he announced his intentions: if he would try for another $5 billion for the California stem cell program?

I have heard Bob speak many times in the thirteen years I worked for him. And every time there is new material: a peek behind the door into tomorrow.

Today was no exception.

He began with a paraphrase of the great Dickens quote from "Tale of Two Cities":

"It was the best of times, it was the worst of times; it was a time of great wisdom, it was a time of great folly; it was the Spring of hope, it was the Winter of despair."

"For the California public, it is critical we seize this historic opportunity for the sake of our children, our wives and husbands, and our parents, to improve and lift conditions which affect us all."

John Zaia (City of Hope photo).

"I am the father of a son who was diagnosed at age eleven with type 1 diabetes. I am the son of a mother who had Alzheimer's, who lost her memory so completely she could not recognize her own family."

"But most importantly, and quite by accident, I am a citizen of California, a state which represents an opportunity to advance stem cell therapy on an unprecedented scale. Why? California is uniquely situated; fifty per cent of America's biomedical research capacity is located in California. If the Golden State was a nation, our biomed sector would lead every country in the world except the United States."

"We know what happened, under the California stem cell agency's new facilities grants, when $271 million was leveraged into $1.2 billion for stem cell research facilities — why? The donor class recognized a unique opportunity; there would be guaranteed research funding for these research institutes, like nowhere else. Congress could not do it, because they cannot bind other Congresses."

"Do I know the position of the current President on stem cell research? I do not. And I am not sure anyone knows his opinion on a week to week basis."

"But I do know the people with whom he has surrounded himself: a vice President, Mike Pence, who opposes embryonic stem cell research; Steve Bannon, an adviser, also adamantly against the research;

Bob Klein (guttmanninitiative.com).

Tom Price, Health and Human Services, with the same religion-based opposition; and Speaker of the House of Representatives' Paul Ryan, sponsor of a personhood bill, to establish a fertilized egg as a person to be protected by the force of law: preventing the use of excess, fertilized IVF eggs, already scheduled to be discarded."

"So, we will not waste our time on how things might be done better in Washington, but rather focus on what we can uniquely do ourselves."

"We have a narrow window of time, which we must take very seriously. We must find a way to sustain and leverage this opportunity."

"If we do, if the polling suggests enough support to provide new funding, $5 billion, then Washington will never again shut the research down."

"There are people in this room who once faced a similar situation with recombinant DNA. Political leaders for religious reasons had protested, saying this was God's territory and none of Man's business; they shut down the research at Harvard, and were trying to stop it everywhere."

"They said, nothing will ever come of this recombinant DNA research, ever, but just one year later, in 1977, came the discovery of artificial insulin, which kept my son alive for many years."

"We must support these scientists today, as they advance the therapies...As my wife Danielle says, "Life is not about waiting for the storm to pass; life is learning to dance in the rain ...""

I had heard Bob wrestle with the giant ideas many times over the years; today's speech was closer than ever before: clear, concise, and irresistible.

But he had not said the magic words; he had not committed to the full-on charge-- to raise another $5 billion ...

Not yet.

39 STEM CELL BATTLES — ON TIMES SQUARE?

One afternoon in 1963, I and another soldier stood on top of the Empire State Building, above the greatest city in the world.

Paris is more beautiful. Los Angeles has movie stars. But, New York City is where everything comes together, city of nations, the meeting place of Earth.

And the heart of it all is Times Square, the most visited spot in the world. At one end is a giant electronic billboard, the Reuters sign — 22 stories high, 7,000 square feet; breaking news, the pulse of the planet.

Remember that billboard, please.

But first, back to the observation deck of the Empire State Building.

A security guard leaned back in his chair against the wall, beside an open door.

Yes, that was the indoor fire escape, he said; and no, we could not climb down it.

So we casually strolled all the way around the observation deck, returning to the guard in his chair who was distracted, by a woman in a bright red dress.

We ducked past him and ran down the metal stairs, into the echoing dark.

The guard hollered furiously, but presently his voice diminished, and was gone.

It was not as exciting as might be thought, running down 86 flights of stairs, each one lit by a naked light bulb. The air was close and still; the walls were caked with dust. A few graffiti told how others had climbed up the indoor fire escape; but nobody had gone down. After a few dozen flights, we began to understand why.

Finally we reached the ground floor, ran to the exit door, pushed on the long brass bar, but nothing happened. It would not open. Stuck!

Climb back up 86 flights of stairs? The doors were locked against theft, and would not open inwards …

We kept shoving on the exit door with various parts of our bodies, and suddenly the wood creaked, yielding half an inch. Encouraged, we pushed and pushed and the door opened out into the light.

Standing on the street, we looked at each other, and burst into laughter.

Like chimney sweeps, our faces were covered with dust, streaked by sweat …

Fast forward half a century.

It's 2016, I was 70 years old, still had the same opinion about New York.

The Empire State stands by its stem cell researchers, two ways.

The New York Stem Cell Foundation (NYSCF) is privately-funded, bringing in about $20 million a year. Roy Geronemus is chairman; Susan Solomon is the legendary fundraiser behind it.

NYSCF has some terrifically worthwhile projects, such as building a bank of 2,500 stem cell lines, trying to represent the genetic diversity of the United States and the world.[1] Imagine having a stem cell line ready to go, one that matched your genetic makeup, and could be used to test drugs on, and maybe hook you up right away with a fix for whatever ails you! That's the goal, and I hope we reach it.

The second program, the New York State Stem Cell Science Program (NYSTEM) is funded by the state, and brings in about two and half times that amount.

In 2006, vision-challenged David Patterson, Lieutenant Governor of New York, was instrumental in the development of NYSTEM. It was originally scheduled to provide $600 million over ten years, $60 million a year. But the funding never came easy. And today, when the progress has been so strong, their money might be cut, or eliminated altogether. What a mistake that would be.

As their recent "Report on Progress" states:

---

[1] http://nyscf.org/

Lt. Governor David Patterson (blackkudos.tumblr.com).

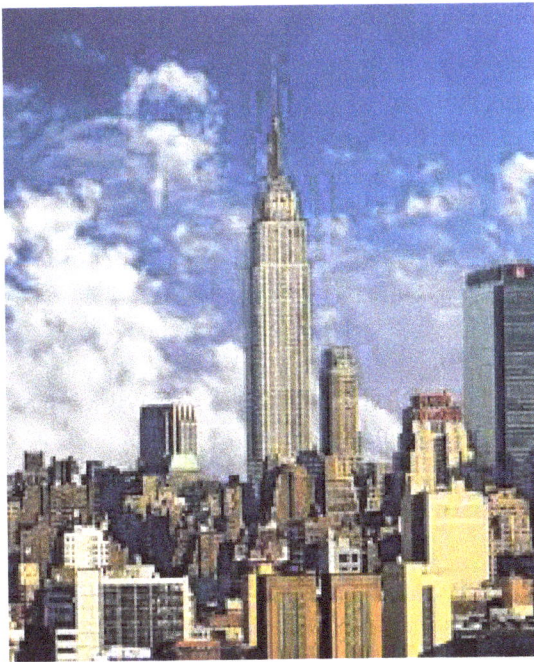

The Empire State Building (www.Britannica.com).

"Our accomplishments are impressive, but more funding is needed to develop and deliver cures and to sustain the vibrant research community that NYSTEM has helped to establish ... to expand the economic benefits of stem cell discoveries for the state's hospitals, universities, research institutions, and taxpayers, and to attract and promote the establishment of successful ... biotech industries ..."

The report was blunt about the problems faced by American scientists:

"... a dramatic decrease in overall funding for biomedical research at the NIH ... unprecedented budget pressures ... Due to inflation, the NIH budget has lost almost 25% of its purchasing power over the last decade ..." (And this is before the Trump Administration's first budget called for an additional 18% reduction! — DR)

NYSTEM funding helped develop several start-up biotech companies, including Blue Rock Therapeutics, with prospects so strong it was able to gather $225 million in base funding. Blue Rock, said Dr. Lorenz Studer, "has two major projects: Parkinson's Disease, and cardiac repair, cooperating with Canada."

New York's research is irreplaceable. Check out just one example from NYSTEM's 76-page report:[2]

"Parkinson's Disease (PD)."

"NYSTEM has been absolutely essential for our ... push toward a first human trial using lab-grown dopamine cells (the ones whose loss causes PD). ..."

"Our project involves basic scientists, engineers, manufacturing specialists, neurologists, surgeons, ethicists, trial experts, and patient advocates ..."

"Thanks to NYSTEM, our group is in an excellent position to be a world-leader in this effort ... to benefit the ... patient population, (and explore) potential economic advantages in ... a multi-billion dollar market in PD pharmaceutics."[3]

That is just one example; NYSTEM has funded literally hundreds of projects, 372 according to the report.

As of this writing, one of their top scientists, Lorenz Studer, has been working nearly two decades to restore the dopamine cells taken away by Parkinson's.

[2]https://stemcell.ny.gov/press-release/nystem-releases-second-strategic-plan
[3]https://www.mskcc.org/research-areas/labs/lorenz-studer

Susan Solomon (twitter.com).

Lorenz Studer (twitter.com).

How are they doing? In a telephone conversation with Dr. Studer, he explained that they had now developed "more than 1,000 doses of the dopamine cell product" and hope to have it in human trials by early next year. The cells have already undergone major testing with mice, rats, and rhesus monkeys.

What were his thoughts on the California stem cell program?

"Visionary ... a bold approach, moving rapidly toward clinical trials ... working closely with the scientists involved ... and stimulating research across the country."

And one more thing: remember that giant electronic billboard in Times Square? Twenty-two stories high, 7,000 square feet?

For 13 seconds, the cover of my small book flashed on that billboard, in front of New York, which means the world.

The cover picture showed then-Governor Arnold Schwarzenegger shaking hands with my son Roman Reed in his wheelchair, and me and Bob Klein looking happy, which was easy, that being the day California loaned CIRM $150 million to get started ...

Here is the picture of that brief instant of notoriety:

"Stem Cell Battles" — on Times Square.[4]

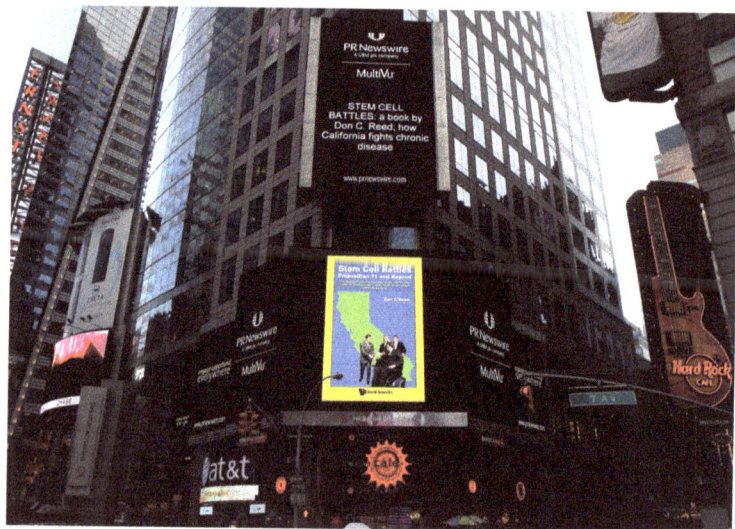

"STEM CELL BATTLES" in Times Square (Huffingtonpost.com).

---

[4] http://www.huffingtonpost.com/don-c-reed/stem-cell-battles-on-time_b_9026044.html

40 ANNETTE, RICHARD PRYOR, AND MULTIPLE SCLEROSIS

"He was born with the gift of laughter, and the sense that the world was mad." — SCARAMOUCHE, by Rafael Sabatini

That line would seem to have been written for comedic genius Richard Pryor. His life was almost unspeakably tragic: raised in a brothel, sexually and physically abused, expelled from school, imprisoned, terrified of performing.[1]

And yet, he gave laughter to the world. Google some videos of him performing, and you will get a stomach ache from laughing. Caution: he uses curse words like punctuation. But funny? Watch him act out the time he stepped into the ring with Muhammed Ali — I have seen it twice, and tears ran down my face both times.

A very different kind of performer was Annette Funicello, star of Walt Disney's original Mickey Mouse Club, and a series of lighthearted movies like Beach Party. Discovered while dancing the role of the Swan Queen in Tschaikowski's Swan Lake at the age of seven, Annette's smile could warm a room.[2]

Like most of America, I watched transfixed when the Mickey Mouse Club was on, hugging my knees in front of our neighbor's 7-inch screen, lasting out each second of the show as long as I could. Annette was special; in my secret heart I really felt she was my friend; that smiling person I would never meet.

Both superstars reached the peak of the entertainment world; beloved by millions.

---

[1] https://en.wikipedia.org/wiki/Richard_Pryor
[2] https://en.wikipedia.org/wiki/Annette_Funicello

Annette Funicello (en.wickipedia.org).

Richard Pryor (en.wickipedia.org).

And then for each the symptoms began.

Fatigue, at first, simple-seeming tiredness, surely nothing a good night's rest could not take care of. But then the limbs would spasm suddenly, or freeze into paralysis.

Worse was to come.

"Blurred or double vision ... thinking problems ... loss of balance ... weakness ... bladder problems ..."

While making a movie sequel with Frankie Avalon, Return to Bikini Beach, Annette had balance trouble. Rumors spread the problem was alcohol: false.

In 1992, Ms. Funicello revealed that she had multiple sclerosis, and established the Annette Funicello Fund for Neurological Disorders. Her husband Glen Holt was a fine man, who stood beside her all the way. But the quality of her life went down and down, until a wheelchair was the only answer. Annette who had once been a dancer now had to be carried.

The lifespan of a person with MS is not appreciably shortened: both stars lived full lives. Richard Pryor died at 65, Annette passed away at 70.

Although symptoms of MS can be eased by medication, the root cause has not been dealt with — until now. MS has to do with the loss of insulation (called myelin) around a human nerve.

But what if new myelin could be put in place, re-insulating the nerves?

In Southern California, an effort to do that is underway. With a California stem cell agency grant, Principal Investigators Craig Walsh at the University of California at Irvine, and Jeanne Loring of Scripps Research Institute are tackling the disease with co-PI Claude Bernard at Monash Institute, Australia.

Note: All California funding must be spent in our home state. But when a scientist in another state or nation can bring his or her own funding, as Dr. Bernard is doing, that means more bang for everybody's buck. Each researcher can leverage the resources and intellect of their collaborator. The scientists win; the patients win.

Dr. Thomas Lane of the University of Utah was also a huge part of the project, "vital to our efforts," said Dr. Walsh. Here is a video of Dr. Loring, giving the "elevator pitch" on how to fight MS.[3]

---

[3]https://www.cirm.ca.gov/our-progress/video/jeanne-loring-scripps-cirm-stem-cell-sciencepitch-multiple-sclerosis

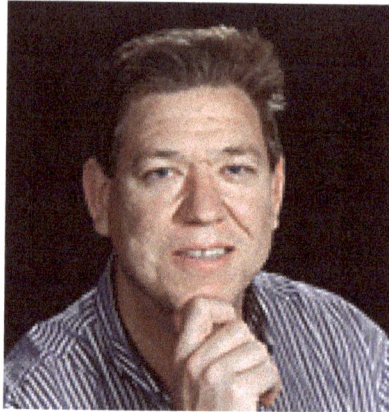

Craig Walsh (national multiple sclerosis society).

Jeanne Loring (scripps.edu).

The mice in the experiment were paralyzed by a virus, in much the same way MS works.

Then, the therapy: "... neural precursor cells (NPC) were derived from human embryonic stem cell line H9. Transplantation of these cells resulted in significant clinical recovery, beginning at 2–3 weeks following transplant."

Paralyzed mice became un-paralyzed. And here is something fascinating. "Despite this striking recovery, these ... [cells] were rapidly rejected."

The transplanted cells were rejected — *but the improvements remained*. This could be great news indeed, and it might apply to other diseases as well.

Consider: one of the problems of transplanting new organs is rejection — the body thinks the transplant is an enemy, and fights it. The doctors have to shut down the immune system, putting the patient on special medication for long periods of time, maybe life. The immune system gone, the person is at risk: a cold can kill.

But if the cells which brought a cure were rejected by the body, but the cure itself remained — there might be no need for long-term, anti-immune medications.

Another surprise: "the work has ... value added' benefits ... Our immune cells can also activate regeneration ..." — (personal communication, Craig Walsh)

How long until people with MS can get well? No way to know.

But because California voters said YES to Proposition 71, the Stem Cells for Research and Cures Act, we have a chance to defeat Multiple Sclerosis.

When that day comes — may it be soon — let us toast the memory of Richard Pryor and Annette Funicello, who gave a smile to the world.

41 MIKE PENCE, AND REPRODUCTIVE SERVITUDE

"Servitude: slavery, a condition in which one lacks the liberty ... to determine one's course of action, or way of life."— Merriam-Webster.

Vice President Mike Pence is a personhood believer, demanding full legal rights for every fertilized human egg. This is no casual whim; the former Governor of Indiana has worked to impose this bizarre ideological belief upon America.

Pence co-authored House Resolution 374, the "Life at Conception Act," which seeks to "implement equal protection ... for the right to life of each born and preborn human person at all stages of life, including the moment of fertilization ..."[1]

Let us be blunt. We are talking about the contents of a tampon. At the "moment of fertilization" described in Pence's law, the fertilized egg, or blastocyst, is a near-invisible dot of tissue, essentially liquid, which women may shed in their monthly cycles. The microscopic blastocyst is gone, without ever being noticed.

This is in no way a "human person." Unless the egg implants in the walls of the womb, it is biologically impossible for it to become a child. Without implantation, there is no pregnancy, and of course, without pregnancy, there can be no abortion.

---

[1] https://www.govtrack.us/congress/bills/112/hr374

But if a fertilized egg was redefined as a person, that could make possible an end run around "Roe v. Wade," the Supreme Court decision which currently protects women's reproductive rights.

The "personhood" idea came about because of something said by Supreme Court Justice Harry Blackmun during the Roe v. Wade court case:

"If this suggestion of personhood is established ... right to life would then be guaranteed ..."

What could it mean, if we officially declared an egg to be a person? Once we enshrine such nonsense into law, all manner of consequences become possible.

A woman's right to control her own body? Gone. If personhood becomes the law of the land, an abortion at any stage will be illegal. Right now, a woman can legally terminate her pregnancy until the point when the fetus becomes "viable," meaning it can live on its own. That means she has roughly 22 weeks to make up her mind, to have the child or not. Under personhood, that cushion of time is gone altogether. Once sperm meets egg, the pregnancy must be carried out in full.

Want to practice birth control? Personhood could limit you to "barrier methods", i.e. condoms or abstinence: no sex at all. "The pill," the most common form of birth control, would almost certainly be illegal because it could now be considered an "abortifacient," a chemical form of abortion, which personhood defines as the murder of a person.

Want to have a baby by the In Vitro Fertilization (IVF) method? Better check with a lawyer. The IVF procedure has allowed more than five million families to have babies. Under personhood, IVF as currently practiced to give a family the best chance of a viable pregnancy and a healthy birth might well become illegal.

In the IVF process, 15 to 20 sperm-and-egg embryos are made, and only the strongest one or two will be implanted. The rest will be either frozen and stored, given away or sold to other couples, or (most often) discarded. If a fertilized egg was legally declared a person, then disposing of even one could be called homicide — and fifteen? Would that not be mass murder?

But wait, there is still more.

Embryonic stem cell research employs blastocysts that would otherwise be thrown away after the IVF procedure. These are "4–5 days old … smaller than the size of the dot over an "i"…[2] This would be illegal too.

Personhood is so extreme Mississippi (arguably the most conservative state in the union) voted it down, 58–42%[3] — and Colorado (where the initiative began) rejected it three times, by a nearly 2–1 margin, 63–37%.

"Personhood"[4] … a word to remember, and a concept to beware.

---

[2] https://www.cirm.ca.gov/about-cirm/cirm-faq#embryos4

[3] https://www.washingtonpost.com/politics/mississippi-anti-abortion-personhood-amendment-fails-at-ballot/

[4] http://www.cosmopolitan.com/politics/news/a31309/personhood-reproductive-rights/

42 MOTORCYCLE WRECKS AND COMPLEX FRACTURES

In 1978, my brother David Reed was driving his motorcycle up a winding mountain road near Santa Cruz, California. A truck came around the other side.

"I smashed into him head-on at 60 mph, flipped over the truck and rolled down a cliff," said David.

He landed in a tree on the hillside, motorcycle on top of him, his right leg shattered.

Fortunately, the truck driver was a decent sort, and did not leave. He called for help, and a stretcher rescue was carried out.

Bone protruded through the fabric of David's jeans in two places. His tibia, fibula and femur were horrifically fractured.

Should David's leg be amputated? I argued in favor of keeping it, in hopes he would be able to walk; also, what if they found a way to cure it later on?

I had no idea what I was asking him to endure.

The pieces of broken bone were removed, cored out, stacked on metal rods, and put back in. After this "healed," the rods were removed, and bone grafts from his hip put in. An experimental device called an osteo-stimulator was used to try and increase bone growth.

His calf had titanium spikes sticking out both sides, like the snout of a sawfish.

When it seemed the worst was over, about two and a half years after the accident, and he had progressed from wheelchair to crutches, he fell on a sidewalk and re-collapsed his right femur. The titanium rod was pushed two and a half inches out of his hip. Surgeons replaced that rod with a new one, shorter now because of the compressed and re-fractured

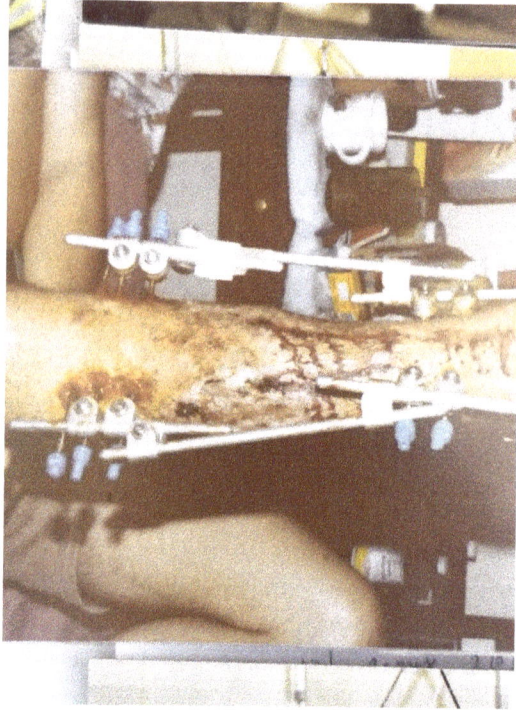

Dave Reed's leg (personal).

femur. He has worn a lift on his right shoe ever since to lessen limping and avoid more back damage.

Attending school on crutches, Dave got two vocational diplomas in electronics so he could work at a desk. He worked as an electronic tech in a factory for six years. When that closed, Dave was walking without a cane. Having been in X-ray school before his accident, he returned to the medical field, and got a degree as a Registered Nurse. He worked as a nurse for 20 years, retiring on disability after falling on concrete while helping another R.N.

Although he had a total 12 surgeries, he says he is glad he kept the leg, because he can walk, as long as no one is in too big a hurry. But the pain continues, and the disability is permanent.

At Cedars-Sinai Medical Center, stem cell scientist Dan Gazit and orthopedic surgeon Hyun Bae are trying to find a better way: to heal

Dan Gazit (bio.csmc.edu).

Hyun Bae (bio.csmc.edu).

bone fractures more quickly, and minimize the pain. These are experts: top folks in their fields.

Dr. Bae graduated from Yale with honors, and is a board-certified orthopedic surgeon at the Los Angeles Spine Institute;

Dr. Gazit heads a research lab at the Department of Surgery and Board of Governors Regenerative Medicine Institute of Cedars-Sinai. He says:

"Segmental bone fractures ... cause great suffering to patients, long term hospitalization, repeated surgeries, loss of working days, and considerable costs to the health system."[1]

Gazit and Bae, of course, can do nothing without a research grant. Almost everywhere, the news is grim about grant availability.

In 2011, osteoporosis (bone loss) research received $179 million in NIH research grants, but only $141 million in 2014: the amount went down $38 million.

Fortunately, the California stem cell program is still going strong.

Dr. Gazit won a "translational" grant from the cell program, and Bae joined him as the clinical lead of the project. Basic science must come first, the foundation of fact on which everything depends: lots of microscopic studies, cells and lab rats. But translational means new diagnostic tools, therapies, medicines, and procedures.

Dr. Gazit sees more than one route to recovery:

"My group has shown that stem cells from human bone marrow, engineered with a bone-forming gene, can lead to complete repair of segmental fractures."

"... Or, we could gene-modify stem cells already in the fracture site.

"We were the first to show, in a rodent model, that a segmental bone defect can be completely repaired by recruitment of stem cells to the detect site followed by direct gene delivery ... we aim to   promote this project to clinical (human) studies."

"Recruitment of stem cells?" In a telephone interview, Dr. Gazit explained that in this approach, no new stem cells would be put into the body.

---

[1] https://www.cirm.ca.gov/our-progress/awards/gene-targeting-endogenous-stem-cells-segmental-bone-fracture-healing

Instead, a scaffold made of collagen (sponge-like material) would be inserted at the wound area. This would attract the body's own stem cells. When enough of them arrived, (and there was a way to keep count of how many) then the genes would be injected. To put the gene into the cells, Dr. Gazit would use an "ultrasound machine that almost every cardiology department has."

"The (inserted) gene would trigger the cells to regenerate the bone …"

"… We will test the method in repairing large bone defects. If successful, we will be able to proceed to FDA approval towards first-in-human trials."

Dave Reed walking (personal).

"... This project could lead to the development of a simple treatment for massive bone loss, which could benefit the citizens of California by reducing loss of workdays, duration of hospital stays, operative costs — "

" — improving quality of life for (those) with complex segmental bone fractures."

People like my brother.

43 EVEN DRACULA GETS ARTHRITIS

In the 1966 movie BILLY THE KID VS. DRACULA, there is a scene that made me cringe, and not for the reasons the film-makers intended.

As Dracula, veteran actor John Carradine lurched toward his prospective victim, and the camera focused on his *hands*.

The swollen-knuckled joints were unmistakable. Arthritis.

The rest of the movie I found myself worrying for the vampire, afraid someone might shake his hand, and squeeze too hard.

Arthritis may not sound like much: perhaps because it is so common, affecting more than 20 million people in America alone. But the pain can be excruciating: it would be classified as torture, if done deliberately.

As mentioned, my wife Gloria and I have the condition in our knees.

For me, it's minor: getting up from the floor is difficult, like a camel rising, rear end first, one awkward joint at a time.

But Gloria suffers.

Inside her knee, where thigh and shin bones meet, there should be a spongy cushion of cartilage. If it was thick and healthy, all would be well.

But that shock-absorbing cushion is worn out, from the wear-and-tear of aging. It is bone on bone as the weight presses down, and every step grinds her joints.

Surgery could re-strengthen the knee with metal and ceramic implants. If the operation went well, Gloria could recover a degree of mobility, after a painful period of recovery. But still that is a foreign object in the body. More surgeries might be required: expensive, invasive, and always with risk.

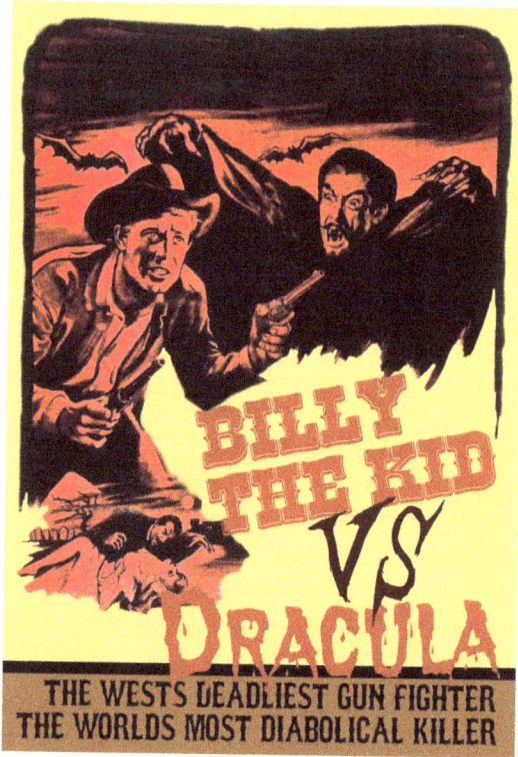

BILLY THE KID VS. DRACULA (Amazon.com).

My Aunt had exactly such a knee surgery. A blood clot broke loose, traveled to her brain, and she died on the operating table.

Gloria has had cortisone shots and gel insertions; but these are temporary fixes, each one helping less than the one before.

But what if we could grow new cartilage, and rebuild the cushion inside the knee?

Denis Evseenko of UCLA is trying to do just that. His grant (RB5-07230) from the California stem cell program is active; the cell of choice is embryonic (ESC).

In a telephone interview, Dr. Evseenko said:

"Recent tests have shown the ESC-derived cartilage cells to be ... functional only in the presence of specific factors ... The Evseenko lab has

Dennis Evseenko (newsroom.ucla.edu).

developed a novel small molecule regulator designed to replace those natural factors. Co-transplantation of the ESC-derived cartilage cells and this rejuvenating molecule is expected to yield the most efficient functional outcomes in patients with arthritis ..." — Denis Evseenko, personal communication.

His work may "delay or even prevent the need for joint replacement procedures ..."

"The work in his lab is now focused on generating these cells on a large scale and testing their ability to repair cartilage injuries ..."

As Dr. Evseenko puts it, "Cartilage restoration (should be) one of the major priorities in medicine."

Toward the end of his life, John Carradine's suffering was severe, and he once described his hands as "arthritic talons."

Yet he made an estimated 500 movies, probably more than any other actor in the world. Some were quite wonderful, like 1940's *Grapes of Wrath*, while others were of lesser quality.

Why did he keep on working? He once described himself as "just a ham," meaning someone who loves the spotlight, but there was more to it than that.

Carradine loved the plays of William Shakespeare.

He would take virtually any film or TV role, however bad, as long as it would help finance his true passion: to perform and direct in a traveling

On right, John Carradine (En.wikipedia.org).

Shakespearean troupe. He would work and work, and whenever he had saved up enough money, he would get his actor friends together, and they would hit the road, bringing the world's greatest playwright to life for people across the country.

It was a speech from Shakespeare which gave Carradine his lucky break in Hollywood. The great director Cecil B. DeMille overheard him reciting HAMLET's soliloquy, and liked the quality of his voice, and hired him.

Listen again to those wonderful words:

"To be, or not to be: that is the question; whether 'tis nobler in the mind to endure the slings and arrows of outrageous fortune — or to take arms against a sea of troubles, and by opposing — end them."

That speech sums up the advocate's choice: shall we passively accept the agony of chronic disease, or fight back with everything we have?

44 TUGBOAT FOR CURE

It was raining as I walked across the UC Berkeley campus: not a flooding downpour, but enough to bring a smile from a water-loving person.

"Go past Sather Gate a quarter mile," a helpful student said, "cross the wooden bridge, turn right, turn left, then right again."

And there it was, the Li Ka Shing Center for Biomedical and Health Sciences: strongly built, five stories high: with two levels just for stem cells.

The Center was one of 12: established by CIRM'S major facilities grants.

What a bargain! All colleges applying had to come up with at least 20% of the grant in matching funds from donors; many gave more.

CIRM contributed a grant of $20 million, matched by a $40 million donation from the great Hong Kong philanthropist Li Ka-shing, and additional funding from the Gordon and Betty Moore Foundation, the Wayne and Gladys Valley Foundation, the Ann and Gordon Getty Foundation, and more new money for California.[1]

Total CIRM costs for all 12 institutes and centers of excellence? $272 million. But with the matching grants, that $272 million was leveraged to $1.1 billion.[2]

Every center had to have a stem cell program with one, two, or three components: basic science (hunting for discoveries), translational (developing new products or therapies), and/or FDA-approved clinical

[1]http://www.berkeley.edu/news/media/releases/2008/05/07_cirmgrant.shtml
[2]https://www.cirm.ca.gov/our-progress/video/cirm-major-facilities-speed-stem-cell-science-and-create-jobs

Jennifer Doudna (UC Berkeley photo).

trials (careful tests on people). UC Berkeley chose to focus on both basic and translational science.

"Basic research is where the magic is," said Lily Mirels, cheerful administrator of the program, "... the desire to understand nature, to discover underlying principles ... Without these, there is nothing to translate... into therapies."

For instance, UCB Professor Jennifer Doudna needed laboratory space to work on her incredible and historic discovery: CRISPR/Cas9, a form of genome editing: potentially of such importance to cure or mitigate genetic diseases that it may well lead to a Nobel Prize.

What does that mean, to edit a genome? Imagine a house blueprint on your computer screen. If you don't like part of it, you cut and paste, editing its design; changing the blueprint alters the house.

The body's blueprint (DNA) may have flawed genes, like those which cause sickle cell disease. Sickle cell anemia warps and stiffens blood cells, until they block the capillaries, bring agony, and shorten a life.

But if we altered the genes, we might modify the body's blueprint, and remove the sickle cell mutation.[3]

Li Ka Shing liked that idea, and contributed $10 million more to develop the Innovative Genomics Initiative (IGI) inside Berkeley's CIRM Center of Excellence.

"Today," said Dr. Mirels, "The IGI is using genome editing to fight sickle cell disease." In a cooperative effort with Benioff Children's Hospital Oakland, scientists including CIRM scholar Mark DeWitt, IGI Scientific Director Jacob Corn, and Mark Walters, Director of the Blood and Marrow Transplant Program, are working together to save children from lives of pain.

It all began at the kitchen table for David Schaffer, Director of the Center.

"My father was a professor of biochemistry," he said, "And my mother was a physician. It was natural to discuss medical research at dinnertime, and I wanted a career like that, developing knowledge and improving people's health."

"The Berkeley Stem Cell Center was actually begun in 2004 by Randy Schekman, who later won the Nobel Prize," said Dr. Schaffer. But the new program needed lab space and equipment, and CIRM made that possible.

The groups of Schaffer and Steven Conolly are fighting Parkinson's disease. They seek to replace destroyed nerve cells, transplant new ones into the brain, and monitor them by a process called Magnetic Particle Imaging (MPI).[4]

Dr. Schaffer is working on spinal cord injury, Alzheimer's, and Parkinson's: seeking to grow, protect and transplant stem cells for relevant therapies. A CIRM scholar, Maroof Adil, brings the energies of youth to the effort.[5]

---

[3]https://blog.cirm.ca.gov/2015/09/18/stem-cell-stories-that-caught-our-eye-new-crispr-fix-for-sickle-cell-disease-saving-saliva-stem-cells-jumping-genes-ipscs-and-lung-stem-cells/
[4]www.magneticinsight.com
[5]https://blog.cirm.ca.gov/2015/12/07/cirm-scholar-spotlight-berkeleys-maroof-adil-on-stem-cell-transplants-for-parkinsons-disease/

David Schaffer (cchem.berkeley.edu).

Xavier Darzacq (cchem.berkeley.edu).

What is happening? First, a change in the "media" (the stuff stem cells are grown in), so that it can be liquid for injection, and gelatinous when stability is wished.

Second, a new way to look inside a body. The Conolly lab is engineering more powerful MPI machines to track the transplanted stem cells.

Happy Note: CIRM scholar Patrick Goodwill, who with Dr. Conolly was "instrumental in developing MPI," is now Chief Technical Officer at Magnetic Insight, a startup company based on this work.

Professor "Call me Xavier" Darzacq and his lab came from Paris, France to UC Berkeley, thanks to a CIRM Research Leadership Award, leveraged by funding from the Siebel Stem Cell Institute.

Advancing work begun by Nobel prize-winner Eric Betzig, Dr. Darzacq is building super-powerful microscopes that can follow a single molecule within a living cell, like tracking one ant in a colony of billions. These allow Darzacq and his collaborator Robert Tjian to study how gene regulation works: turned on and off like light switches inside stem cells: healing wounds and regenerating tissues.

How does Dr. Darzacq feel about the California stem cell program?

"I was talking to the mechanic working on my car," he said, "and I asked him, did he know about the stem cell program California began with Prop 71?"

Li Ka Shing Center for Biomedical Sciences (Jenny Cutting picture, UC Berkeley).

"When he said he vaguely remembered a controversial proposition, I told him that he should be so proud, because it has now become a statewide enterprise to cure diseases using stem cells. It creates work and business in California and provides hope for patients, and your vote made it possible."

As Dr. Mirels walked me out, I looked back at the beautiful building. Something about the shape of it tugged at my consciousness. And then it occurred to me.

The Li Ka Shing Center had the outline of a tugboat: those work horses of the sea which rescue far larger ships, pulling them to safety through the storm.

45 WHEELCHAIR WARRIORS, TAKE BACK YOUR RIGHTS!

Question: Should voters with a disability (and their families) take part in the elections?

Answer: Only if we want to win.

We have gigantic numbers on our side. According to the United States Census, "Nearly one in five people have a disability ... 56.7 million people — 19 percent of the population."[1]

That is one of the biggest constituencies on the planet! My son Roman Reed drives a wheelchair. He will vote, as will every eligible member of his family.

As Roman puts it: "Of course we are going to register and vote. That is our right and our duty. But it is not enough just to vote. We must also reach out to our friends in the community to add to our strength, helping each other."

Enable the disabled! Wheelchair warriors, spread the word. Ask your friends if they are registered to vote. Do you have your ride to the voting booth figured out? Plan it early. Don't lose your vote because suddenly no driver is available. If you have room for somebody else, offer a ride.

Voters with a disability may have trouble getting to the polls. If you are blind or paralyzed, for example, driving may not be an option.

But, if you are disabled, a minority person, a student, a low-income citizen, or a senior like myself, remember there are politicians who do **NOT** want you to vote. In more than two dozen states, Republicans have

---

[1] https://www.census.gov/newsroom/facts-for-features/2015/cb15-ff10.html

passed new laws to make it difficult: requiring, for instance, government-issued ID cards (like drivers' licenses) which non-drivers may not have.[2]

The most vicious voter-suppression trick is called Interstate Crosscheck "which purged 1.1 million Americans from the voter rolls" in the 2016 election.[3] How it works is simple and deadly. Voters with names similar to those in another state. For example, James Edward Harris Jr. of Richmond, Virginia (and) … James R. Harris (no Jr.) … are subject to getting scrubbed off the voter rolls. When they show up, they are given a provisional ballot, and told they must bring documentation as to who they are to the Secretary of State's office. Most people won't go to that extra trouble (they may have to work) and that cancels their vote.

The Cross-Check system must be fought. Meanwhile, do not let any obstacle stop you. In every state, there are ways you can vote, and people who will help you.

Here is a great phone number for general voting information: **1-866-OUR-VOTE**. I called them, and they were very helpful. They will ask you what state you want information about, and then transfer you to someone who can help.

You may choose an absentee ballot. Don't know how to get one? The office of your Secretary of State will help you; that is their job.

Or, contact the League of Women Voters, who will help you.[4]

Vote early. Not sure how to register? Another useful source is: Vote411.org.

Above all, vote. Remember Ed Roberts, who attended college at UC Berkeley while in a portable iron lung which he invented? One of the great organizers of voters with a disability, he did not want charity or pity. He wanted to be at the table when decisions are made. Because of his efforts, we now have curb cuts for wheel chairs and centers for independent living. We honor his memory best by voting.

---

[2] http://www.huffingtonpost.com/don-c-reed/voter-suppression-steals-_b_13069770. html http://www.gregpalast.com/trump-picks-al-capone-vote-rigging-investigate-federal-voter-fraud/

[3] http://www.gregpalast.com/trump-picks-al-capone-vote-rigging-investigate-federal-voter-fraud/

[4] http://lwv.org/member-resources/voter-registration

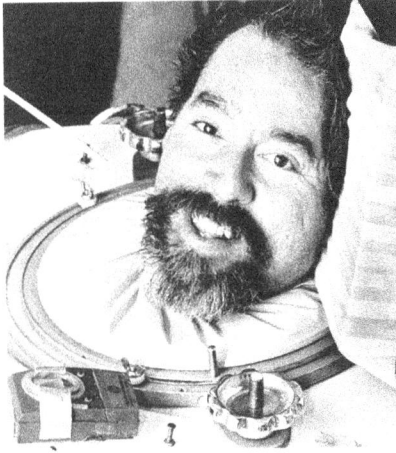

Ed Roberts (www.Smithsonianmag.com).

Do not disenfranchise yourself. Let your voice be heard and your vote be counted; enable the disabled!

"Democracy never sleeps. Every citizen must be vigilant: for our families, our country, our future ..." — Bob Klein, July 8, 2017

Remember in November ...

46 SICKLE CELL DISEASE VS. STEM CELL AGENCY

Inside the veins of an African-American child, there are red blood cells: round and soft, doing their job, keeping him or her alive. But what if those cells hardened and changed shape, curving into the letter "C," hard as a cornflake?

Sickle Cell Anemia.

Those capillaries clog, in a "crisis." Excruciating pain, like broken glass in the veins, a crisis may last an hour or a day, and the damage is ongoing.

"By 20 years of age, about 15% of children with Sickle Cell suffer major strokes ... by 40 years of age, almost half have central nervous system damage ... cognitive dysfunction (mental problems) ... damage to lungs and kidneys ... frequent hospitalizations ... (all leading to) an early death ..."[1]

Poverty makes things worse: African-Americans are almost three times more likely than whites to be poor,[2] and accordingly lack decent medical insurance.

Without proper medical advice, Sickle Cell Disease (SCD) sufferers may not know even basic care: like the vital importance of drinking lots of water. The more water they drink, the less often their blood cells will clog the capillaries.

---

[1] https://www.cirm.ca.gov/our-progress/disease-information/sickle-cell-disease-fact-sheet
[2] http://kff.org/other/state-indicator/poverty-rate-by-raceethnicity/?

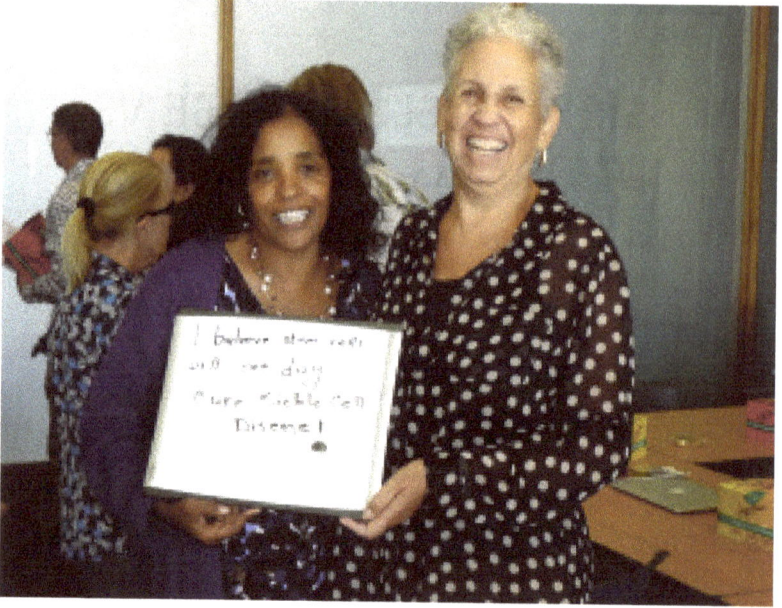

Adrienne Shapiro and Nancy Rene (blog.cirm.ca.gov).

Ted Love (endpts.com).

"Another effective treatment is a medication called hydroxyurea, which reduces crises by 50% and death by 40%, but most adults are not treated."

The problem is getting worse. For a person with sickle cell anemia ...

"... the average life expectancy has (lessened) from 42 years in 1995 to 39 today."

But doesn't the Affordable Care Act (ACA, or "Obamacare") help?

Yes, if your state has it. The Affordable Care Act (ACA) brought reasonably-priced medical care to millions of the poor and middle class. If someone on "Obamacare" develops SCD, medical care and advice is available to them. This is a wonderful accomplishment, for which history will thank President Obama.

Politics, unfortunately, has gotten in the way.

"As originally enacted, the ACA required states to expand Medicaid eligibility to (families) ... with incomes (close to) the federal poverty level ... However, a 2012 Supreme Court ruling made it optional for states to expand Medicaid eligibility ..."

As of this writing, 21 state governors (all Republicans) have "opted out" of ACA: denying affordable medical care to millions of low-income Americans: including all too many families of color.[3]

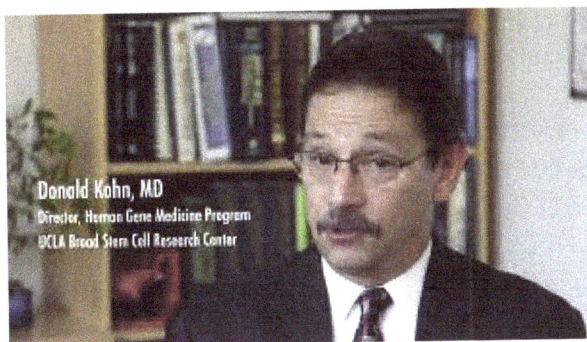

Donald Kohn (UCLA Broad Stem Cell Research Center).

---

[3] http://money.cnn.com/2013/07/01/news/economy/medicaid-expansion-states/

And, as this is written, the Republican party is trying furiously to end Obamacare, denying medical care to millions ... Such cruelty is beyond understanding.

But access to care is only part of the problem; we need to have therapies waiting, actual cures to make people well, not just maintain them in their misery.

Bert Lubin is a member of the California Stem Cell Agency's board of directors.

If you live in Oakland, California, you might know Dr. Lubin, who for more than 36 years has been working to save children's lives from sickle cell disease.

"At the Children's Hospital of Oakland, Dr. Lubin began the Sibling Donor Cord Blood Program, offered to families across the United States who have a child with a blood disorder such as sickle cell anemia ... and who are expecting another child. Following the birth of a healthy child, (his/her cord blood) ... is harvested. Because cord blood is enriched with blood-forming stem cells, it is cryopreserved (frozen) and can be later used for transplantation. A number of lives have been saved following transplantation with cord blood units collected in this program ..."

Another champion is Dr. Ted Love, who recently retired from the board of the California stem cell program.

Dr. Love is one of the most genuinely charming people you will ever meet, and he has a way of calming down arguments that is amazing to see. When disagreements on the stem cell board got hot and heavy, he would sum up both sides with a gentle voice, allowing problems to be settled amicably. When I told him he would be greatly missed, he said he wanted to dedicate his life to finding a cure for SCD.

How is Dr. Love doing?

"Helmed by Bay Area biotech veteran Ted Love, Global Blood Therapeutics is developing treatments for sickle cell disease ... a genetic blood disorder that in the U.S. affects 1 in 365 African-Americans ..."[4]

---

[4]http://www.xconomy.com/san-francisco/2015/07/08/tackling-sickle-cell-global-blood-therapeutics-joins-ipo-queue/

The third man is Dr. Donald Kohn of the University of California at Los Angeles (UCLA), recipient of funding from the California stem cell program, and famous for his work with the Bubble Baby Syndrome.

His method?

"... (Take) some of the patient's (bone) marrow and then use gene therapy methods to correct the ... defect in the blood stem cells before (putting) them back into the patient ..."

What would this do?

"... correction of the sickle mutation in adult bone marrow would allow for permanent production of normal, non-sickle red blood cells ... These advances will have direct and immediate applications to enhance current medical therapies ..."

Will this information be kept secret, or be shared with other researchers?

"All scientific findings and biomedical materials produced from our studies will be publicly available to non-profit and academic organizations in California ..."

"The first clinical trial of stem cell gene therapy has begun at UCLA and there are great hopes a new therapy will emerge." — Dr. Kohn, personal communication.

Can our scientists end this deadly disease? California intends to find out.

Because black lives matter.

47 DWIGHT CLARK, "THE CATCH," AND A.L.S.

One of the greatest moments in the history of sports was the 1981 NFC Championships between the Dallas Cowboys and the San Francisco 49ers.

With just 58 seconds to play in the 4th quarter, Joe Montana was about to be buried by Ed "Too-Tall" Jones. Montana faked his pass, causing Jones to leap prematurely. Unable to see past the huge man, Joe fired a bullet to where he hoped Dwight Clark would be. The pass was high, but Clark leaped for it, stretching, stretching … and sunk his fingers into the ball, coming down into athletic history.

I was there, and "the Catch" looked easy. Only afterwards, when it was replayed millions of times, could it be seen what an athletic miracle it was …

Just days ago, Dwight Clark announced that he had ALS, Amyotrophic Lateral Sclerosis, motor neuron disease, called Lou Gehrig's disease in America.

Another world class athlete, Lou Gehrig, the great New York Yankee baseball player, was called the "Iron Horse" because he always stayed in top condition, never failing to start a game. Inexplicably, he began losing control of the ball. Throwing, catching and hitting — his coordination was destroyed. He trained harder, saw doctors; but nothing helped.[1]

Few know how he spent his last months, serving as a probation officer, working with ex-cons, trying to give them the second chance he would not have himself. When Lou Gehrig died, America knew about ALS motor neuron disease.

---

[1] http://www.espn.com/nfl/story/_/id/18955466/san-francisco-49ers-legend-dwight-clark-diagnosed-als

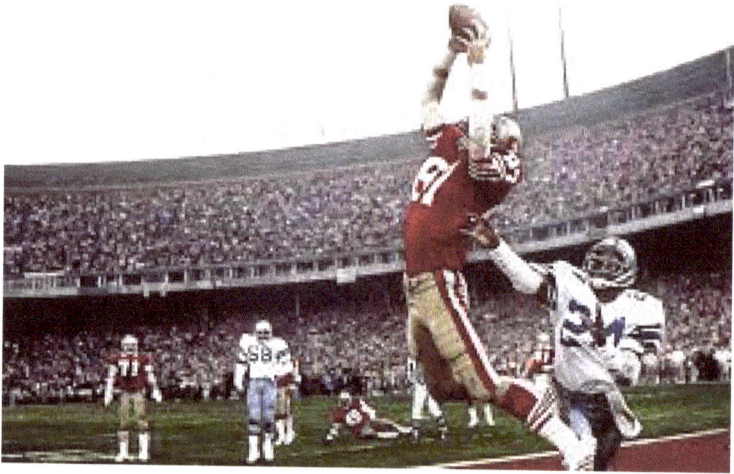

Dwight "the Catch" Clark (Nydailynews.com).

ALS kills the nerve cells which control motion. Gradually, over a period of several years, the body loses its ability to move: even to swallow, and, finally, to breathe.

"I lost two sons to ALS,"[2] said Diane Winokur, ICOC board member, and a champion patient advocate for research to cure this devastating disease. Fragile-looking, but with a will of iron, Ms. Winokur has worked decades, and now at last she has seen the scientists bring a pinpoint of light to what was darkness before.

CIRM scientists are seeking to more clearly "understand the origin of the disease" and what causes (the death of) motor neurons ..."[3] They must know the enemy, cells literally oozing toxin, damaging the nerve cells which control motion.

And the path to cure? One promising possibility: "(Our) scientists learned how to take ... cells and turn them into motor neurons (for possible replacement)."[4]

---

[2]http://www.huffingtonpost.com/don-c-reed/diseaseaweek-challenge-1-_b_7455178.html
[3]http://quest.mda.org/news/mda-study-reveals-cost-illness-als-dmd-mmd
[4]https://www.cirm.ca.gov/our-progress/disease-information/amyotrophic-lateral-sclerosis-als-fact-sheet

If new and healthy nerve cells are introduced into the body, they may replace the damaged ones, and reverse the downward spiral to paralysis and death.

None of this happens without money. The California stem cell program has provided $55 million in research funding, toward the cure of ALS. Such funding makes possible the work of top scientists (and their lab teams!) champions like: Larry Goldstein, Bin Chen, Clive Svendsen, Eugene Yeo, Bennett Novitch, Ying Liu, Steven Finkbeiner, Zack Jerome, and more. They work at world-class research institutions like: the University of California at San Diego, Cedars-Sinai Medical Center, the J. David Gladstone Institutes, and the Salk Institute for Biological Studies — all institutes benefiting from a CIRM relationship.

Dwight Clark gave joy to uncountable millions of sports fans. We all wish him well, him and everyone who suffers a disease labeled "incurable." I hope he knows, he is not fighting alone.

FLASH! On July 20, 2017, the California stem cell program "board voted to invest $15.9 million in a Phase 3 clinical trial run by Brainstorm Cell Therapeutics ... using mesenchymal stem cells ... from the patient's own bone marrow ... modified ... to boost production of ... proteins (which) help support and protect cells destroyed by ALS ... a therapy called NurOwn ..." — Kevin McCormack, Communications Director, CIRM.

"These two trials that CIRM is now funding represent breakthrough moments for me and for everyone touched by ALS ..." — Diane Winokur.[5]

---

[5]https://blog.cirm.ca.gov/2017/07/20/stem-cell-agency-funds-phase-3-clinical-trial-for-lou-gehrigs-disease

48 A FRIEND IS LOST

Yelling "FIRE, FIRE!" my six-year-old brother came scrambling down the hill toward our back yard. Behind him rose a plume of black smoke. The campfire he had started had not stayed small, spreading to the dry grass.

Neighbors reacted, charging up the hill.

When we got to the top, a wall of crackling flame was closing in on a large wooden shed, and behind that a small neat house with nobody home.

My father uncoiled a garden hose; I found a stack of old throw rugs. Everybody attacked the fire, stopping it a couple feet short of the wooden shed.

Curious to know what we had saved, I poked my head inside. In the quiet shadows were rows of stacked cans marked "FLAMMABLE" — paint thinner ...

If we had not stopped the burning when we did, I might not be writing this column.

Like the stages of the fire, some diseases start out small and then explode.

In 2003, a woman named Theresa Blanda was diagnosed with polycythemia vera (PV), a disease where the blood thickens, producing too many cells.

Complications ensued. Like the fire, her condition grew worse, like scar tissue developing in the bone marrow. Her limbs ached, her knees swelled. She had difficulty breathing when lying down, because her spleen had enlarged to the size of a football. She had to quit work.

Catriona Jamieson and Theresa Blanda (www.cirm.ca.gov/our-progress/stories-hope-leukemia).

And waiting just ahead, like the paint-thinner explosion if we had not stopped it? A blast of cells, acute myeloid leukemia (AML), the disease which killed my sister.

"You couldn't even breathe my way, or I'd bruise," said Theresa Blanda. "I did not think I was going to make it ..."

But at Theresa's side was blood specialist and stem cell expert Dr. Catriona "Cat" Jamieson, Deputy Director of the Sanford Stem Cell Clinical Center. Partially funded by the California stem cell agency, Dr. Catriona Jamieson had made a discovery that made a tremendous impact on Theresa's health.

It was a mutation in the blood-forming stem cells ... a gene called JAK2.

Like the lighting of a match, that mutation was the start of everything bad. But what if the flame could be put out? What if that mutation could be inhibited?

Aided by a complicated funding arrangement involving academic and biomed entities (including a 501 c-3 organization called San Diego CONNECT) Dr. Jamieson was able to identify an already existing drug, a JAK 2 inhibitor, that could be re-purposed to control the deadly disease.

Ms. Blanda was admitted into a clinical trial, where she received the inhibitor, developed by a San Diego biotech company, TargeGen, Inc. She began to to breathe normally, as her spleen shrunk back. She could climb stairs again. As the knee swelling went down, so did the back pain, and she returned to work ....[1]

For eight years everything seemed fine, and then the FDA suspended the Phase 2 clinical trials. Some of the other patients had developed negative side effects.

As veteran science writer Bradley J. Fikes put it, "Without the drug, her health began to regress. After having a seizure, she was rushed to the intensive care unit."

Dr. Jamieson and her team did not give up. Racing against time, with Ms. Blanda's life hanging in the balance, they developed another inhibitor, which the FDA put an emergency rush on, and approved, but it was too late.

At her funeral, her family thanked the California stem cell program, for having given eight more years of life to Theresa Blanda.

In addition to the California stem cell agency and the Jamieson team, numerous organizations cooperated: the Dana-Farber Cancer Institute, the Moore's Cancer Center at UC San Diego, the Mayo Clinic, the FDA, a parallel study by Harvard Medical School, San Diego Connect, and a biomed outfit called TargeGen.

May their efforts be remembered, scientific neighbors trying to save lives.

---

[1] http://www.sandiegouniontribune.com/business/biotech/sdut-ucsd-moores-patients-blanda-jamieson-2015feb21-htmlstory.html

49 DYING IN DOONESBURY, FIGHTING BACK AT UCD

For nearly half a century, Gary Trudeau's cartoon strip *Doonesbury* has been a part of our lives, bringing laughter, tears and the itch of thought.[1]

Having followed the cartoon story through its many book collections, I think of the Doonesbury family almost as people: for instance, two characters; Joanie Caucus, and Andy Lippincott.

When Ms. Caucus had a crush on handsome Andy Lippincott, I was glad for her; when he turned out to be gay, I was startled. Then I forgot about his character.

But when Andy came back into the story again, he was wearing a hospital gown. He had lost much weight, his hair was gone — he had contracted HIV/AIDS.

Even so, I was sure everything would be all right. When Joanie Caucus brought him the long-awaited Beach Boys album "Pet Sounds," it was a gift to a living person, with a future ahead of him.

And then, one stunning cartoon panel. Andy was lying on his side, his back to us

Offstage, Joanie called to him ... "Andy, are you okay? It freaks me out when you tune out like that!" But Andy Lippincott just lay still.

It was the first time I emotionally realized people died of AIDS.

Today, there are pills which keep HIV/AIDS patients alive. But they still have shortened lives, more than ten years less than average, and are at increased risk of diseases like cancer and stroke. In addition, they face medical costs of roughly $2 million over a lifetime of the pills.

---

[1]https://en.wikipedia.org/wiki/List_of_Doonesbury_characters

Doonesbury's Andy Lippincott (gerryco23wordpress.com).

How many Americans have HIV, which may develop into AIDS? 1.2 million.[2]

But the California stem cell program will still bring the fight.

Check out their cooperation with:

Calimmune, Inc., the company begun by Ronald Mitsuyasu of UCLA, and strengthened by Nobel prize winner David Baltimore;

John Zaia at City of Hope;

Sangamo Biosciences, Inc.;

Paula Cannon at USC and the UC Davis team.

The University of California at Davis is a powerhouse in stem cell research: located near Sacramento, California's State capitol. Dr. Jan Nolta leads the UC Davis Institute for Regenerative Cures, nearly 150 scientists working on a variety of stem cell projects.

On July 23, 2015, CIRM okayed an $8.5 million grant to UC Davis, "to test gene modified   stem cells in patients, and then monitor and analyze their effectiveness (against) HIV."[3]

Led by Mehrdad Abedi and Joseph Anderson, as well as research veterans Gerhard Bauer, Richard Pollard, and Xiao-Dong Li, the team hopes to make:

"... a one-time treatment, with the possibility of controlling ... HIV ... by eliminating the reservoirs of HIV in patients (which are) responsible for persistence of the disease," said Dr. Abedi.

---

[2] https://www.cdc.gov/hiv/basics/statistics.html
[3] http://www.ucdmc.ucdavis.edu/publish/news/newsroom/10204

Is that possible? A "one-time treatment"... like going to the doctor just one time, and then you are well? Wow.

In addition to the funding, CIRM also provided a "clinical advisory panel," said Dr. C. Randall Mills, President and CEO of CIRM. "These advisory panels consist of subject matter experts, CIRM science officers, and a patient advocate who works with the researchers ..."[4]

When AIDS is finally defeated, it will be a group victory. Hundreds of scientists, thousands of advocates, millions of taxpayers will deserve applause.

And a tip of the hat to cartoonist/satirist Gary Trudeau, who took us into the hospital room of a cartoon character, to show us why we fight.

[4] https://www.ucdmc.ucdavis.edu/cancer/Newsroom/feature_articles/072415-Oncologist_lead_stemcell_trial.html

50 THE MAN WITH THE AUTOGRAPHED BASEBALL

How do you repay someone who saves your life?

In the junior high school where I taught, a woman was eating lunch, alone in the Teacher's room. Suddenly a piece of meat went down the wrong way and got stuck. She could not breathe. She got up, staggered out into the hallway, and saw the vice Principal approaching. She could not speak, but she pointed to her throat.

He turned her around, made a fist of his left hand, covered it with his right, and yanked it upward at the base of her ribcage. A piece of steak shot out and landed on the hallway floor ... She breathed again. He had saved her life.

The next day, by way of a thank you, she brought him a plate of cookies.

Brenden Whittaker is a small quiet man, who reminds me of Kit Carson, the great explorer. Whittaker, a former Boy Scout, used to love to hike the Appalachian mountain trail, until a chronic disease took over his life.[1]

Chronic granulomatous disease sent him to the hospital "hundreds of times," and cost him "parts of his lungs and liver."

But Dr. Donald Kohn and Caroline Kuo of UCLA was performing a clinical trial,[1] which "removed Brenden's blood stem cells, genetically re-engineered them to correct the mutation," and put them back.[2]

---

[1] https://blog.cirm.ca.gov/2017/01/19/stem-cell-profiles-in-courage-brenden-whittaker/
[2] https://stemcell.ucla.edu/news/new-clinical-trial-funded

Brenden Whittaker (BW personal collection).

That is an important three-step process, a pattern that I suspect will become wide-spread. Take some of the blood out, fix it, and put it back. The healthy cells fight the bad ones — hopefully defeat them — and the patient is a patient no more.

Today, Brenden Whittaker does not have chronic granulomatous disease.

The staff at CIRM autographed a baseball for the man whose life had been saved by medical science, and how is he repaying the scientists who gave him his life?

In the best of all possible ways: now that he knows he will have enough lifetime to do it, Brenden Whittaker is going back to school — to become a doctor.

51 THE GORILLA GYNECOLOGIST RETURNS

Whatever scientific accomplishments Dr. Bertha Chen, MD, achieves (and as we'll see, they are substantial), to me she will always be the "Gorilla Gynecologist."

Remember Koko, the gorilla who "spoke" with sign language? She was having trouble conceiving, and so Dr. Chen examined her, and found her healthy; the reproductive problem turned out to be with the male. Asked what the gorilla's personal equipment was like, Dr. Chen replied:

"Very similar to human females, only somewhat larger."

What was it like, working so closely with such a gigantically powerful creature? Did she have a picture of the two together?

"Unfortunately not, I was too scared to be thinking of a photo. It's frightening to have to insert an instrument into the vagina of such a large animal! She could easily have flattened me if she closed her legs while I was doing the exam."

— Bertha Chen, Stanford, personal communication.

But, while that incident enshrines her forever in the history of medical adventure, one of Dr. Chen's research innovations might literally save your life.

First, remember how a therapy developed for one disease may be helpful with another? Hold that thought. First, her usual work, then the stunning surprise.

Bertha Chen (med.stanford.edu).

Chen's specialty is Urinary Incontinence, (UI) which affects both men and women.[1] She and Renee Reijo-Pera worked for years on ways to control UI.

The mechanics are simple: The bladder is an internal bag for urine storage. The sphincter is an outlet. When the bladder is full, the sphincter opens, ideally under the owner's control, and lets the urine out.

Smooth-muscle cells keep the sphincter closed. If they become weak or deficient, accidents will happen, sometimes including continual leakage.

Women commonly lose control of their bladders because of damage to the tissues surrounding the birth canal when having a baby. Men skip that difficulty but have prostates, which can go wrong. I can vouch for this, unfortunately.

Both genders are affected by age, when those smooth muscle cells cease to function in the sphincter — a lot of incontinence! More paper is used for adult diapers than for babies. The sheer recycling challenge is daunting, 3–5 diapers a day for seven million women in just California? Add half again that amount for men, and that's about 30 million diapers a day, in just one state.

---

[1] https://www.cirm.ca.gov/our-progress/people/bertha-chen

The financial costs of incontinence are estimated at $25 billion a year.[2]

The greater expense is human tragedy: women institutionalized for this condition. The danger is in falling. If an elderly woman falls, fragile bones may break. If she wakes up at night and rushes to the restroom, and trips, her pelvis may shatter.

Surgery can be done to treat urinary incontinence, but it is risky and less effective in older people. The most common non-surgical technique is "bulking", putting collagen, fat, or ground-up bone into the sphincter. But this only narrows the opening and does not restore muscle control.

Is there a better way?

Dr. Chen works with smooth muscle stem cells, which close the bladder's sphincter. When you are hurrying to the restroom, you know exactly where those muscles are, as you try to prevent leakage by internally squeezing them.

Age decreases these stem cells, until they are not enough to function.

But what if we could make new stem cells for anyone who needed them?

Smooth muscle stem cells might be injected into the sphincter, returning the clenching muscles to a youthful vigor.

That could improve the quality of life for millions.

But (here it comes) there may also be another and extraordinary benefit.

Think of the biggest artery in your body, the aorta, from the left part of your heart, down into the abdomen. Sometimes, aortas begin to thin, weaken, and bulge out — an aortic aneurism.[3]

If the aorta breaks, the blood can gush like a fire hose into your body cavity, and you die in a couple of minutes.

Or maybe not.

Imagine if that was you: and a CT ("cat") scan showed a developing weakness in the aorta, like a bulging spot on an air-filled bicycle tire.

---

[2]http://www.mipsnet.org/single-post/2015/08/25/Economic-costs-of-urinary-incontinence

[3]http://www.mayoclinic.org/diseases-conditions/abdominal-aortic-aneurysm/symptoms-causes/dxc-20197861

You could have a major operation, try and patch it up before it blows open.

Or, if Dr. Chen's theory is correct, smooth muscle stem cells might simply be injected into the bloodstream (probably through catheters in the big artery at the groin), to strengthen the walls of the damaged aorta.

To make the cells match the patient, Dr. Chen would use the Yamanaka method, making Induced Pluripotent stem cells from the patient's own skin.[4]

And to think all this began when the California stem cell program gave a research grant to a scientist trying to help us oldsters use the bathroom voluntarily!

---

[4] https://www.ncbi.nlm.nih.gov/pubmed/19030024

52 WRESTLING THE INVISIBLE ENEMY

Long ago in Junior High, I was crossing the track field after school. I was by myself, but up ahead was another teenager.

Suddenly, he fell down on the grass, violently; his body flung itself around, for no apparent reason, like wrestling somebody who was not there.

I ran up beside him, but had no idea what to do. I just stood there, watching, until he lay still.

"It's okay," he said when he calmed down, "No big deal, happens all the time."

He got up, and went his way, seemingly unharmed.

It was epilepsy, the disease Supreme Court Justice John Roberts may have.[1]

In North America, an estimated 2.9 million children and adults have active epilepsy. An average adult has a 10% chance of having one or more seizures.[2] For some, what happens is subtle; they may stare at nothing for a few seconds, losing contact with the world. For others, seizures are so severe that some must wear football helmets to protect themselves. Sometimes, the helmets break.

If you observe something similar, doing nothing is generally the right thing to do. Do not try to hold them still; do not put anything between their jaws. They will not swallow their tongue; the biggest danger is a rock or something they may bang their head on. Talk softly to them, and

---

[1] https://www.washingtonpost.com/national/health-science/why-john-bryson-and-john-roberts-should-talk-about-their-seizures/2012/06/18/
[2] https://www.cdc.gov/epilepsy/basics/fast-facts.htm

Arturo Alvarez-Buylla (neurosurgery.ucsf.edu)

stay, be a comforting presence until the seizure passes. If the seizure lasts more than 5 minutes, call 911.[3]

For most, drugs control the condition. But for 20–30%, the pharmacy offers little assistance. Surgery may be required.

Brain surgery? To physically remove part of one's brain? The brain is who we are, the headquarters of our lives: coordinating bodily functions, enjoyments and exertions — the brain is involved in virtually everything we do, conscious or not.

Part of the epilepsy problem is what happens between individual nerve cells, the neurons. If there is too much nerve excitement, things go wrong. A healthy brain deploys cells to calm that excitement, "inhibiting" it.

Listen to Dr. Arturo Alvarez-Buylla of UC San Francisco, personal interview:

"Many (nerve) disorders show an imbalance between excitation and inhibition …"

When the nervous system becomes too excited, signals are disrupted. The brain may throw the body into seizure.

---

[3]https://www.cdc.gov/epilepsy/basics/first-aid.htm

Usually, drugs can calm the nerve circuits. But this may not work, which is why cell therapy may be needed.

With a group of UCSF colleagues and experts (authorities like Drs. John Rubenstein, Arnold Kriegstein, Scott Baraban, Michael Stryker and Allan Basbaum), Alvarez-Buylla is using healthy cells which normally come from a part of the brain called the (sorry!) "medial ganglionic eminence (MGE)."

They are making precursor cells, the in-between stage. (First they are "blank check" embryonic stem cells, then precursors, then young neurons, before becoming the final stage, the mature inhibitory neuron.)

When these precursor cells are grafted into the brain, they (amazingly) migrate to where they are needed, and become inhibitory cells. They link up with nerve circuits, and hopefully repair what is not working.

By way of full disclosure, Dr. Alvarez-Buylla stated that he is a co-founder of Neurona Therapeutics, and sits on that company's scientific advisory board. This company is a start-up, built around the idea of using MGE cells in various ways.

But back to the story.

As you recall, over-excited nerve cells in the brain must be calmed down, and MGE cells have been recruited for the job. They are differentiated properly, and transplanted into the brain, where they "swim" to where they are needed.

Will it help fight epilepsy?

"Recent studies ... demonstrate that MGE cells grafted onto an animal model of epilepsy can significantly decrease the number and severity of seizures."[4]

But wait (as they say on TV), there's still more! And it's big time ...

"Other 'proof-of-principle' studies suggest transplanted MGE cells can be effective in the treatment of Parkinson's disease, and in the control of chronic pain."[5]

Parkinson's? Chronic pain? My ears perked up at that. To fight Parkinson's disease was astonishing enough, but to perhaps also end the need for painkillers?

---

[4]https://www.ncbi.nlm.nih.gov/pmc/articles/PMC4056344/
[5]http://journals.plos.org/plosone/article?id=10.1371/journal.pone.0061956

We've heard the stories. A soldier gets wounded in battle in Afghanistan, comes home, and takes opioids for the pain … and becomes addicted.

As a kid, I jumped off a playground swing (showing off for a couple of girls) and broke both arms. After the casts were put on, I tried to go to sleep, but could not. The pain was astonishing, and it just went on and on, not easing up until the next day. I was young and healed swiftly, of course. But what if I had to endure that agony on a permanent basis?

Dr. Alvarez-Bullya's efforts could be world-changing; a stem cell attempt to ease epilepsy, perhaps leading to control of Parkinson's, and ending chronic pain?

Interviewing him, however, was a bit of a struggle. Modest to a fault, the Mexican-American scientist persisted in pointing out the accomplishments of his colleagues.

"Dr. Basbaum is the world authority on pain, you should really talk to him … Dr. Kriegstein's work is outstanding … Dr. Rubenstein's basic science is crucial …"

It did not occur to him to mention that his 30-plus years of effort had won him numerous top-level awards, including one, the Prince of Asturias Award, which cited his and his co-recipients' findings as "among the most important in neurobiology, changing the way we understand the brain..."[6]

---

[6] https://translate.google.com/translate?hl=en&sl=es&u=https://es.wikipedia. org/wiki/Arturo_%25C3%2581lvarez-Buylla&prev=search

53 TWO WARRIORS NAMED JOAN

When Joan of Arc was captured, it was because she was literally too brave. When she led her soldiers into battle, she went first, wounded several times. When retreating, she was always among the last, helping the foot soldiers to safety.

On the day her luck ran out, attempting to relieve the siege of Compiegne, she and her brother Pierre were last to leave the field, with several hundred English soldiers just behind. Still she might have been all right. The castle just ahead of her was well fortified. But Compiegne's ruler lost his nerve. He ordered the drawbridge pulled up, leaving Joan and her brother Pierre outside to be caught.

Imprisoned, she refused to promise she would not try to escape, and attempted several times. Once, she jumped 60 feet, and was knocked unconscious by the fall. Another time she reportedly tricked a guard, shoving him into her cell, and locking the door. She might have escaped, but she tried to rescue her brother, and was caught.

During the trial, (charged with wearing men's clothes!) she spoke courageously.

"I die through you, Bishop!" she said to Cauchon, may his name be remembered in infamy. But the game was rigged; the English wanted her dead.

Tied to the stake, abandoned by the Church she served and the King she made, Joan's last speech was to pray for forgiveness — for herself and those about to set the fire.

Joan of Arc (Istock by Getty Images).

Mercy of a quick death was denied her. It was common to stab a condemned person in the heart, so they were dead before the flames bit in. But this was not done.

The only kindness shown to her came from an English soldier, one of her former enemies. She asked for a cross to die with, and the soldier made one for her, from pieces of a broken stick. She kissed it and placed it next to her heart. Then her arms were tied around the stake, and the wood piled up.

Eyewitnesses said she did not scream when the flames attacked, but only cried out "Jesu ... Jesu ... Jesu ..." from within the fire.

But when she was gone, the king she had crowned regained his courage, ignored his weak advisers, and went to battle. Joan's vision was realized as the French rose up, and drove the English out forever.

No one can match the courage of Saint Joan.

But there are folks who come close.

Over the years I have worked on various out-of-state efforts to provide funding for stem cell research, and to protect the scientific freedom of researchers.

I learned to always contact local Parkinson's folks.

Other groups might promise to get back to you, someday, maybe, when they got around to it. They would have to meet with various committees, possibly next year, but they really could not get involved in anything "political" …

But the Parkinson's people just asked: how could they help?

In Arizona, a woman named Rayilyn Lee Brown would always come through, if I needed a letter of support. She had a keen-edged intellect, words that rang out strong and clear. (She and another Parkinson's patient, Dianne Wyshack, worked unpaid for many years, gathering and sharing information about stem cell research, helping the struggle however they could. Ms. Wyshack passed away not long ago, and we are all diminished by her loss.)

Rayilyn Lee? Once, someone opposed our research on religious grounds, saying scientists were "playing God" with human lives, and called the researchers Nazis.

Ray said:

"Blastocysts are not people. Those of us who support embryonic stem cell research do not believe a blastocyst is a person any more than an acorn is an oak tree. Every seed does not become a plant nor does every blastocyst become a person unless the condition of implantation (is) met … Otherwise, it is a few microscopic undifferentiated human cells that, if left alone in a petri dish, will NEVER become anything.

"Are scientists "playing God" by manipulating cells? … Many people believe God gave man the gift of intelligence to heal suffering …"

— Rayilyn Brown, Director AZNPF

Arizona Chapter National Parkinson Foundation

The world takes pride in another Parkinson's fighter, Michael J. Fox, who turned his struggle with PD into a national effort so far his organization, the Michael J. Fox Foundation (MJFF) has raised more than half a <u>billion</u> dollars for research.

Recently, MJFF merged with the Parkinson's Action Network (PAN) …

Joan Samuelson (c-Span.org).

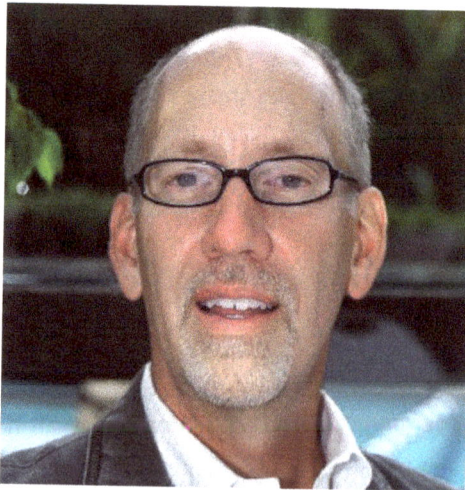

David R. Higgins (California stem cell report by David Jensen).

And where did PAN come from?

A lawyer named Joan, Joan Samuelson was diagnosed with Parkinson's disease. But she would not passively accept it.

As Amy Comstock-Rick, one of her colleagues (later CEO of PAN), put it, "Joan threw her might against the condition," and founded PAN.

Joan worked as Chair for years, and then was chosen as the Parkinson's representative on the board of directors of the Prop 71-derived stem cell effort.

I knew her through her years of effort as an ICOC board member, from the very beginning, when she worked tirelessly to organize the fledgling stem cell program. She was always the one who would ask the difficult questions, partly because she was a lawyer by training, but mainly because she was just Joan.

Toward the end of her multi-year involvement with CIRM, mobility was plainly becoming more and more difficult. As the condition progressed, she required an attendant. Sometimes, he would literally have to pick her up and carry her.

At last a fall at home forced her to cease her active involvement in CIRM.

So, how would you like to be the person chosen to "replace" Ms. Samuelson?

This was the chore of David R. Higgins, who now represents Parkinson's on the ICOC board. He has PD himself, and knows what Joan's effort had cost her.

When asked what it was like to follow in Joan's footsteps, he said it was:

"Terrifying. Joan's shoes were big enough to fit a whole family," he said in a recent interview.

He wanted to go visit her, to express his appreciation for the courage of her life. But her health was not up to it, she said.

Joan of Arc, Joan Samuelson: two warriors named Joan.

54 AN END TO HEROISM?

A mother and her son sat in the living room: she in a chair by the wall, he in a cranked-up hospital bed.

He could speak, and move his arms, turn his head from side to side, but that was all. His mother was equally trapped, by love for her son.

The only escape was the TV set on the wall, and the unconsciousness of sleep. If the young man needed to leave the bed, it required a mechanical hoist. Once I saw Mom wince and clutch her back, clearly in constant pain from having to turn him over. A great fear for caregivers is that they will "throw their back out" and become incapacitated, unable to care for their loved one.

Suddenly they squabbled, viciously, as though a visitor was not in the room. She remembered there was company, and said, apologetically:

"We fight a lot. Cabin fever I guess." She seemed embarrassed, but she did not need to be. Wheelchair rage and attendant fatigue are inescapable.

She was a hero, giving up her own life to provide for his. She must help him with all the activities of daily life we take for granted.

My paralyzed son Roman, fortunately, is very independent. A tetraplegic, he is paralyzed from the shoulders down, but is still extraordinarily active.

Several times a week, I go over to his house to help with the necessaries. But other than that, he is up and gone, maintaining a schedule few could imitate.

Nothing is easy. When he "transfers" from bed to chair, it is an act of herculean exertion. He shifts his bodyweight, and scoots over. Roman is a big man, 6'4", 235 pounds. Consider what "transfer" means. It's almost like a gymnast doing an "L-sit," weight supported on his hands, legs straight out before. Roman has to do that heroic exertion without the use of leg or stomach muscles.

It is at the edge of impossible, and he does it every day, just to get out of bed.

May there come a time, when heroism is not required for the simple chores of everyday life.

55 MESSAGE FROM THE MIDDLE KINGDOM

One of the most spectacular failures of my life was a ten-year effort to learn Chinese — with no teacher. I worked at it 45 minutes a day on my own, using audio cassettes, CDs and textbooks, mumbling Mandarin on the freeway.

In the "Middle Kingdom" (according to legend, China is halfway between Earth and Heaven, therefore the Middle Kingdom) scientists are working hard to cure paralysis, and I wanted to speak to them in their own language.

But maybe I started too late, because the new language did not stick.

I can say one key sentence: "*Wo shuo de bu hao*," which means "I speak very badly," which sums it up. Fortunately, most Chinese scientists speak fluent English. Also, they are patient with my pathetic attempts at communication, invariably responding with a polite "*Ni shuo de hen hao!*", you speak very well — a courteous falsehood.

Like all modern nations, China is very interested in regenerative medicine, and has spent (by 2016) approximately 3.5 billion ren min bi (roughly $512 million U.S.) on stem cell research.[1]

A recent claim to have possibly "reversed Alzheimer's" in mice has the field abuzz. According to a paper in Stem Cell Reports, mice with an Alzheimer's-like brain condition were returned to near-normal status. Dr. Naihe Jing is a name to remember, for all who hope for an AD memory-loss cure.[2]

[1] https://burnstrauma.biomedcentral.com/articles/10.1186/s41038-016-0046-8
[2] http://www.cell.com/stem-cell-reports/references/S2213-6711(15)00272-6

Naihe Jing (The Stem Cellar).

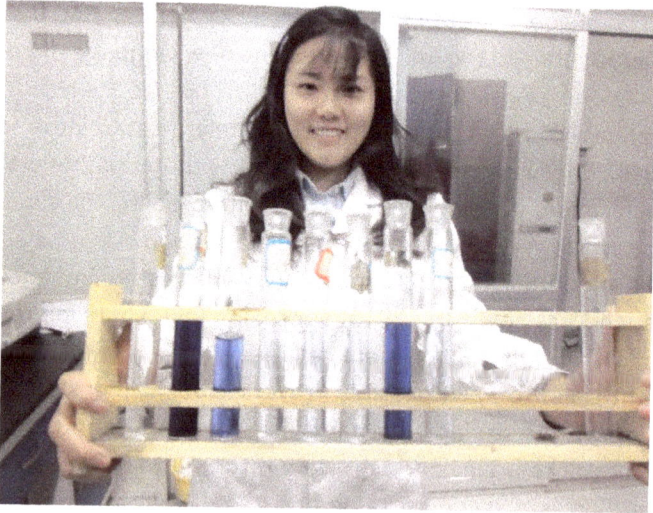

Li Fei (personal collection).

Doug Sipp (personal).

But there is a behind-the-scenes struggle China is engaged in right now: a battle which, sooner or later, every nation must face.

Stem cell tourism.

Stem cell tourism means visiting a foreign country to purchase medical care. At its best, it can open up the world's medical resources; at its worst, it is a nightmare.

There is nothing wrong with the idea itself. For instance, Mexico prohibits human embryonic stem cell (hESC) research, for religious reasons. If I lived in that country, and had a disease or disability which hESC might help, and America had an FDA-approved therapy for it, I would be across that border in a heartbeat.

I have personally participated in stem cell tourism with the critical distinction that the medical product I obtained was FDA-approved for a clinical trial.

When Roman was first paralyzed, there was a clinical trial going on which might have helped him. It involved a substance called Sygen, basically dried-up cow brains. I tried to get Roman into the trials, but was denied, told I had missed the deadline for inclusion by one hour.

However, with the aid of Congressman Fortney "Pete" Stark, who interceded with the FDA, I obtained a compassionate use permit, allowing us to try the medication. But I had to find it first. After locating the inventor, Dr. Fred Geisler, who said it still might work, even past the initial time limit, he pointed me to a source for the drug, Switzerland. Then I had to find a doctor willing to prescribe an experimental drug —, and purchased several cases of Sygen. It was not cheap: $35 an injection, and Roman got two a day. Our family went deep in debt, trying for cure.

Roman recovered the use of his triceps, the pushing muscles on the backs of his arms. Whether it was the Sygen or the grueling rehab that helped, I cannot say. But Roman now drives an adapted van, lifts weights, and lives pretty much on his own.

Would I do it again? In a heartbeat.

But again, the Sygen was in FDA-approved clinical trials and no surgery was required. It would be a very different risk altogether to have a surgeon open Roman's back and insert some unknown substance into his living spinal cord.

Your condition can always get worse.

A few months ago, in Florida, three women tried an adult stem cell remedy to improve their vision. Fat cells were removed from their stomachs, injected into their eyes, and all three women went blind.

Important: they paid $5,000 each to the clinic which did the procedure. To me, that means it was a for-profit procedure, not a legitimate clinical trial.[3]

In America we have the Food and Drug Administration (FDA), which is deserving of respect, but is increasingly understaffed and underfunded.

China has a similar agency, the China Food and Drug Agency (CFDA) and recently adopted new regulations to guide and control stem cell research.

How are they dealing with stem cell tourism? Three key points:[4]

1. Only Tier three hospitals (the best in China) are legally allowed to perform regenerative medicine clinical trials.
2. All trials must be pre-approved by authorities: the Ministry of Health (MOH) and the Chinese Food and Drug Administration (CFDA). Human reproductive cloning is prohibited.

---

[3] http://www.cell.com/cell-stem-cell/pdf/S1934-5909(11)00583-2.pdf
[4] https://www.nytimes.com/2017/03/15/health/eyes-stem-cells-injections.html?_r=0

3. No one may pay for inclusion on a clinical trial.

So: good hospitals, no unapproved trials, and the element of greed is lessened.

If these requirements were uniformly enforced, it could end the scourge of stem cell tourism not only in China, but across the rest of the world as well.

This matters to me personally. My son Roman Reed has worked many years to fund research which might lead to a cure for spinal cord injury. In time, I hope to see his company, StemRemedium, Inc., with a branch in China.

Roman recently hired a communications director, cross-border specialist, Li Fei of ZDG — Zhongguancun Development Group — to investigate such possibilities. Li Fei is perhaps best known in America for her work in helping "scientists from China, the UK, France, Spain, Hungary, Greece and Italy organize a joint research program on boosting the yield of rice through the use of fungi that grows on the root system of the crop ... a major benefit to lower income areas of the world to help increase their rice yields and reduce the likelihood of starvation." — Wei Luo, Chief Operating Officer, ZGC Capital Corporation.

Nothing is easy, of course. The lure of big money is enticing, and there are always those who try to get around the law. International regulations expert Doug Sipp of Riken, Japan, adds: "There is also some uncertainty regarding whether the new laws apply equally to clinics that are entirely privately funded, or to China's large network of military hospitals." — (DS)

But this is a solid framework on which to build.[5] China is also reportedly working with the International Society of Stem Cell Researchers (ISSCR) to continue to improve their regulatory structure.

A way to end stem cell tourism would be enlightenment indeed: a wonderful message from the middle kingdom.

For more: Rosemann, Achim and Sleeboom-Faulkner, Margaret (2016) New regulation for clinical stem cell research in China: expected impact and challenges for implementation. Regenerative Medicine, 11.[1] pp. 5–9. ISSN 1746-0751.

---

[5]http://www.nature.com/news/china-announces-stem-cell-rules-1.18252

FLASH! As this is written, June, 2017, a potentially spectacular set of clinical trials has just begun, under the leadership of Dr. Qi Zhou, Deputy Dean of the University of Chinese Academy of Sciences.

At the First Affiliated Hospital of Zhengzhou University, in Henan Province, Professor Zhou will be guiding embryonic stem cell clinical trials on both Parkinson's Disease and Macular Degeneration blindness.

Every procedure will be carried out under China's new and careful regulations.

As with all science, there is no guarantee of success. But anyone interested in stem cell research should follow this effort. Parkinson's Disease and blindness are the initial targets; but other Chinese scientists are reportedly gearing up for spinal cord injury efforts and much more.

Our best wishes go to Dr. Qi Zhou at the outset of this double set of clinical trials, built on his many years of previous work. His bio includes the following:

"Prof. Zhou's research includes the mechanisms of cell reprogramming, differentiation and de-differentiation, cell fate control, obtaining and maintenance of stem cell pluripotency, etc. In addition, he establishes animal models, including mouse, rat, porcine, primate and cell models of studying human disease. He is also dedicated to promoting the application of stem cells to clinical use." — Zhou, Qi: Institute of Chinese Academy of Sciences.

Gan xi bao yan jiu fei chang hao! (Which means, I hope: Stem cells! Very Good!)

56 SCIENTISTS AND THE UNDOCUMENTED

Racism robs us all: not only those who are denied access to a decent life, but also we who are cut off from their strength, skill and talent.

A previous chapter told how Mexican-American Dr. Arturo Alvarez-Buylla is fighting for stem cell therapies for Parkinson's disease, pain-relief, and epilepsy.

But what if he had been born on the wrong side of a wall, grew up without the advantages of education, and was stuck in field labor all his life?

In high school, I worked one summer in the fields, side by side with people who were almost certainly undocumented workers.

Picking strawberries looked so easy: just bend down to the ground, and twist each berry gently off its stalk, being careful not to bruise it, or you don't get paid.

Each row was a quarter-mile long, and completing it earned one dollar. When I finished the first row (it took three hours) my back was screaming.

But an old brown woman called to me.

"You no make money that way — too slow! Come sit in the shade, I show you how pull stems."

As her fingers flew, yanking the green stems, setting the berries into plastic boxes, she told me about her life: how every year she and her family followed the crops.

One man told me he was building a house: working the season brought him a few dollars, he could buy some bricks and mortar, extend the wall of his future home.

They all sent money home, remittances, to keep the old ones in the family alive. A fine on those hard-earned remittances, I would later learn, was how Kris Kobach advised Donald Trump to make Mexico pay for the wall.[1]

Where did they sleep?

She pointed to a big packing crate, lying on its side. I peered within, saw the burlap sacks they used for blankets, they and the children …

For me, it was a summer job, one I would never wish to do again.

For them, undocumented Mexican-Americans, it was life.

They worked a land their ancestors once owned. One third of the United States had been Mexico — California, Nevada, New Mexico, Utah, Colorado, and Texas. But the Mexican-American War, "Invasion from the North" changed the boundaries. As one observer put it, "We did not cross the border — the border crossed us!"

They were not, of course, the only minority to suffer.

The Chinese endured an actual law targeting them: the Chinese Exclusion Act: the only time in U.S. history when an entire ethnic group was banned. Signed into law by President Chester A. Arthur in 1882, the Exclusion Act prohibited Chinese from entering the United States. The act was renewed for decades, separating countless families. Men who had come over to work in the mines, expecting to bring their wives later, were denied their companionship.[2]

People without the proper papers …

All this came together for me in late January, 2015.

I received an e-mail from a scientist friend, struggling to stay in the country legally, asking for a letter of recommendation. Naturally, I wrote it, reproduced below, the names redacted.

January 6, 2015

To Whom It May Concern:

This letter is in enthusiastic support of the petition of Dr. W., applying for permanent residency in the United States.

Champion scientists like W. are incredibly valuable. He recently identified a novel approach to treating multiple sclerosis using stem cells which significantly reduced the disease's severity in an animal model.

---

[1] http://www.kansascity.com/news/politics-government/article63164887.html
[2] https://en.wikipedia.org/wiki/Chinese_Exclusion_Act

Dr. W. received his MD in a top medical school in China, and became a physician in endocrinology. But he found that medical practice can only treat a limited number of patients. This motivated him to come to the US to obtain his Ph.D degree and conduct stem cell research in the US.

Dr. W. founded a startup company aiming to treat large populations of autoimmune disease patients with the new stem cell technology. He obtained support from a state stem cell grant and private investors. Presently, he is working on pre-clinical studies following FDA guidelines to obtain a clinical trial approval, which could happen in 1-2 years. Dr. W. is not only working on multiple sclerosis, but in collaboration with scientists across the country, he is also working on other diseases such as spinal cord injury, inflammatory bowel disease, diabetes, arthritis etc. I was excited to learn that Dr. W. will soon open a new subsidiary in California to expand his R&D. He is considered one of the top scientists in this field in the world, an expert in translating basic stem cell research into real clinical products. Dr. W.'s research should have an enormous impact on the physical health and economic wellbeing of the entire U. S. population.

I am glad to recommend Dr. W. as a person of exceptional ability. His research achievements show a rare combination of talent, enthusiasm and scientific acumen that will enable him to bring relief to those who suffer. Our country will benefit by allowing him to permanently pursue his important work in the US.

Please let me know if I can be of further assistance in regard to this extremely worthwhile individual.

Sincerely,

Don C. Reed

I am happy to report that for Dr. W., the system worked. He got his green card, expanded his company, and continues his stem cell research today.

But what if he had been denied? America would have lost his contribution.

57 THE GIRL, THE BANDIT, AND WOMEN IN SCIENCE

In 19th century Southern China, according to legend, there lived a young woman named Yimm Wing Chun. She lived with her father, making and selling bean curd for a living. Hearing of her beauty, a local bandit warlord rode into her village with his soldiers, and demanded that Wing Chun come out.[1]

When she did, he looked her over, nodded, and told her he was going to give her a great honor, allowing her the privilege of becoming his concubine.

Wing Chun said: "I thank you for this honor, but I am promised to another."

"Not a problem," he responded, "Tell me who it is and I will have him killed."

Just then, a Buddhist nun named Ng Mui ran up to Wing Chun, and whispered in her ear.

Wing Chun spoke: "Auntie reminds me that I can only marry a man who can defeat me in hand to hand combat."

"I hurt people for a living," the warlord reminded her, "Is that what you want?"

Ng Mui whispered again.

"You are very strong; I ask one year to prepare," said Wing Chun.

By now the whole village was listening.

"I will give you three months," said the warlord, "But I will return in that time. If you are not here, I will burn this village to the ground."

---

[1] http://wingchunmasters.com/history

Cheng Pei-Pei as Ng Mui on left (www.wing-tsun.it).

He left, and Wing Chun went to live with Ng Mui, training in a new style of fighting, designed so a weaker man (or woman) could fight somebody stronger. For example, punches and kicks had no visible windup, more difficult to block.

Time passed. The warlord returned with a red wedding palanquin for her.

He got off his horse, approached Wing Chun, who waited for him with her left hand extended, palm up, and her right hand open and up, guarding her throat.

He laughed, approached, extended his left hand to grab her breast.

Instead of flinching back as he expected, Wing Chun turned over her left hand on top of his, yanked him off balance — and her right elbow broke his nose.

She did not give him time to recover, but came quickly in. Her right foot collapsed his forward knee, then lifted hard, so that her knee smashed his face.

He went down, then fought his way back up — into her waiting fists, fired in rapid succession. The affair continued similarly, until the decision was reached.

His friends carried him home in the red palanquin.

Ng Mui named the new style after her student. When Wing Chun married, she taught her husband, who taught someone else, down through several generations, reaching Leung Bik, who taught Ip Man, who taught Bruce Lee.[2]

I don't know if the legend is true or not, but Wing Chun itself (the style of fighting) is intelligent, elegant, and practical.

I practice Wing Chun, and can do a (very poor) version of its three empty handsets, reputedly developed for the battle between the girl and the warlord. In my garage is an odd-looking wooden device, the wooden man or Mok Jung, which allows students to practice their moves. Every so often I spend ten minutes on it, and still know a little of the Mok Jung form.

The most satisfying film version of the Wing Chun legend is Kung Fu Wing Chun with Pak Ching as an athletic heroine, and the elegant Kira Hui as Ng Mui. The big budget *Wing Chun* with *James Bond's* Michelle Yeoh was fun, but has little real Wing Chun in it. The best movies of Wing Chun techniques in action are the modern Ip Man series with Donnie Yuen. PRODIGAL SON with Sammo Hung, Frankie Chan and Lam Ching Ying has several wonderful Wing Chun fights. But my favorite is "STRANGER FROM SHAOLIN, with Wong Hang Sao. Her training sequences show the beauty and connected unity of the forms.

Key lesson? The elder woman mentors the younger. This is essential, if women are to take their rightful place in science.

For example, my daughter Desiree worked her way to the top of her field, as the Athletic Director of the University of Nevada at Las Vegas: imagine the value of her counsel to a young woman starting off in sports administration.

For the field of biomedicine, I turned to Dr. Anne-Marie Duliege, (AD), to share her thoughts, how she might encourage girls and young women to learn biomed.

Dr. Duliege is a senior executive in the biopharmaceutical industry and a member of the Independent Citizens' Oversight Committee (ICOC),

---

[2]https://en.wikipedia.org/wiki/Ip_Man

Desiree Reed, Athletic Director, University of Nevada at Las Vegas (KNTV.com).

CIRM's governing board. For purposes of the interview, INT is short for Interviewer, i.e. me.

INT: Dr. Duliege, in the Chinese legend of Yimm Wing Chun, a young woman defeats a bandit warlord in hand-to-hand combat — thanks to the training and advice of an elderly Buddhist nun. As a professional in the worlds of business and biomed, would you share your story and perhaps some thoughts to young women considering a biomed career? Where are you in your career right now?

AD: I am a pediatrician and I focused my career on drug and vaccine development. I am now at a biopharmaceutical company in the Bay Area, focusing on novel targets to address autoimmune diseases. A few years ago, I was appointed by the California State Controller to serve on the Independent Citizens Oversight Committee (ICOC) as a representative of the pharmaceutical industry. (Note: Only four members of the ICOC can be drawn from the biomed or pharmaceutical industry, thereby insuring the dominant majority will be patient advocates and scientists.)

INT: What are your objectives as a member of the ICOC?

AD: The ICOC is composed of a group of patient advocates, biotech executives, and research leaders. Our objective is to support stem-cell based research and to advance potential treatments and therapies. Our main responsibilities include making the final decisions of all grant and loan awards, overseeing operations of the CIRM, (California Institute for Regenerative Medicine), and ensuring that all decisions are held to the highest scientific, medical and ethical standards.

Anne-Marie Duliege and scientist daughters: Camille Ezran, and Marie Ezran (personal).

INT: What motivated you to seek a career in biomed?

AD: As far as I can remember, I always wanted to alleviate human suffering. This sprang from the early childhood years I spent in Madagascar. After medical school in Paris, I sought opportunities to return to Africa and work there as a pediatrician. In remote areas of Cameroon and Tunisia I was exposed first-hand to the limitations of clinical care in settings of extreme poverty: the harsh realities that political, cultural, environmental and economic constraints impose on healthcare.

Later, as part of a nonprofit organization, I traveled to Ethiopia to help set up a nutrition camp during the 1985 famine. A few months later, when access to food had improved, I helped change the camp to a basic healthcare clinic. These experiences had a transformative impact on my life as a young physician.

INT: So travel was an important part of your career?

AD: My advice to anyone contemplating a career in medicine or biologic science is to travel early on, extensively, and as far from home as possible. When we get exposed to medical and cultural environments that are different from those we grew up in, our learning curve becomes multidimensional and exponential.

INT: How did you first put your education to work?

AD: Around that time, the HIV epidemic was starting to spread around the world. HIV infection was a death sentence. One of the most tragic aspects was the transmission of HIV from mother to child. We did not understand it, and therefore could not prevent it.

At the first international HIV conference on this topic, I met Professor Art Amman, a renowned pediatrician and professor who had described the first cases of HIV infection in children. He invited me to join his team in the clinical research department of Genentech.

INT: Genentech of course is a world-renowned biomed business, one of the true pioneers in the field.

AD: Working at Genentech alongside scientists like Art Amman and others triggered in me a passion to develop new drugs or vaccines, get them approved by regulatory agencies like the FDA, and make them available to patients.

Katherine Reed climbs Mt. Fremont (Terri Reed photo)

We first focused on testing experimental treatments in HIV-infected pregnant women, trying to prevent passing on the disease to their newborns. Later on we expanded our efforts to the vast and challenging field of HIV vaccine development.

INT: And today?

AD: Fast-forward 30 years. Although an HIV vaccine is still not available, considerable progress has been made: substantial reductions in the risks of contracting an HIV infection and seeing it progress to full-blown AIDS. And, the neonatal transmission of HIV can be completely prevented, if the appropriate treatment is given to an HIV-infected pregnant woman at the time of the birth of her child, as well as relatively inexpensive treatments to the child for six weeks.

INT: Your career is varied!

AD: Over the years, I have worked in large and small biotech companies. I am now the Chief Medical Officer at Rigel Pharmaceuticals, Inc., a biopharmaceutical company focusing on novel targets to address autoimmune diseases.

INT: You must have developed quite a network of friends and business contacts?

AD: Colleagues have helped, challenged, and inspired me throughout my career.

INT: How important is volunteer work?

AD: I maintain a voluntary clinical appointment at Stanford's Lucile Packard Children's Hospital. This provides me with a balance between clinical care and research. In addition to serving on the ICOC board of directors, I remain committed to the field of HIV prevention as a board member of the AIDS Vaccine Advocacy Coalition. I am also Chairwoman of Global Strategies, whose mission is to improve access to healthcare for women and children in some of the most vulnerable regions of Africa.

INT: The next generation of young scientists would have a lot to learn from you.

AD: I try to give back to society by seeking opportunities to coach, encourage, mentor and help young professionals, especially young women, as they consider a career in medicine or the life sciences. I hope that my passion for biomedical sciences, international health, clinical research and medicine can inspire and encourage them, just as others have helped me along the road.

INT: Any last thoughts for young scientists?

AD: The journey is far from easy. My own path has at times meandered back and forth, and has certainly been paved by unexpected turns of events, major disappointments, and some exceptional satisfactions. Keep the motto, "Patients First!" forefront in your mind. Seek advice from colleagues and mentors. Travel abroad, keep an open mind about new medical techniques and new fields, and have the courage to stand up after a fall. And always remember, your two best friends along the journey: persistence and optimism.

INT: Thank you, Dr. Duliege, both for your comments, and your ongoing career.

58 THE GREATEST PROPOSAL

When Joshua Francois proposed to my daughter Desiree, he did so in the classiest manner imaginable.

He began by asking Desiree's brother Roman for his blessing on the marriage. Roman first threatened him with extreme physical violence, if he ever was mean to his sister, and then beamed and wished them both the very best.

Then the young suitor came over to Gloria's and my house, and formally requested our daughter's hand in marriage.

For me, it was easy. The first time I met Josh, I whispered to Gloria, "This is the one she will marry." I had that father's instinct: here was a man who would die for my daughter. Gloria burst into tears, which is, to put it mildly, not usual for her.

And the next evening …

Gloria and I were all dressed up at Papillon's, the fanciest restaurant in town.

In the door came Desiree: tall, beautiful, and a trifle irritated. She had not known they were going out to a restaurant and was not dressed for it; and there was Josh: happy and nervous at the same time.

At each of our places was a bowl of red tomato soup. And on the surface of each bowl was a word, written in yellow mustard. Josh asked Desiree to read the soup:

"Will" … (move to next person) "you" … (continue around the table) … "marry" … (gasp) "me"?

Then it was Desiree's turn for the waterworks.

Advocates: Bob Klein and friends (personal).

After dinner, Josh took her home to change, then out in a limousine for dessert and dancing; they were married three months later in the Monterey Bay Aquarium …

I can think of only one proposal which might top that: for the Golden State.

Will California … replenish our stem cell program with $5 billion dollars?

The need is great; so must be our response.

America has nearly one person in two with a chronic disease: that's 45%, or 133,000,000 people with a long-lasting or incurable disease.[1]

We face the modern-day equivalent of the Black Plague, the disease which devastated Europe in the Dark Ages, wiping out one-third of the

---

[1] http://www.fightchronicdisease.org/sites/default/files/docs/GrowingCrisisof ChronicDiseaseintheUSfactsheet_81009.pdf

people. We may not (yet) have the wagons rolling through the streets every night, and the drivers calling, "Bring out your dead!" But the percentage of sick people is close.

How much do we spend on chronic disease?

In 2017, as this is written, medical care will reach $3.35 trillion.[2]

Of that, 86% goes to chronic disease — roughly $2.9 trillion.[3]

Compare that number to last year's federal income taxes for individuals and corporations: $2 trillion.[4]

And last year's installment of the National Debt? Roughly $700 billion.[5]

One year of chronic disease ($2.9 trillion) costs America more than the federal income tax and the national debt ($2.7 trillion) put together.

And those are just the financials. What about the people behind that mountain of medical expense? Those millions of individuals are all somebody's loved ones, perhaps members of your family or mine. The cost of their suffering? Incalculable.

That is why California fights: to develop cures. And we have already begun to win.

But we cannot rush it. We need what Dr. Duliege called her "two best friends: optimism and determination."

Think of the farmer and a new field. He/she does not wave a magic wand and presto, there is food. Boulders must be removed, and the stumps of dead trees. Then comes plowing, and planting of seeds, plus watering, fertilizing — it all takes time. But if every step is taken and the ground is fertile, the seeds will break open underground, and pale shoots reach up toward the sun. The crops will come: inevitably.

That is where we are today. California planted the seeds of cure research, and the first rewards are already here: lives have been saved, sight has been restored to the blind: motion to the paralyzed. We have

---

[2] http://www.pbs.org/newshour/rundown/new-peak-us-health-care-spending-10345-per-person/

[3] https://www.cdc.gov/chronicdisease/resources/publications/aag/pdf/2015/nccdphp-aag.pdf

[4] https://www.cbo.gov/topics/taxes

[5] http://www.washingtonexaminer.com/cbo-budget-deficit-to-rise-to-693-billion/article/2627452

the momentum; we need to keep going. What a mistake it would be to quit now!

We need a Prop 71 Part Two: what Bob Klein has talked about — a $5 billion replenishment to the California stem cell program.

It will be the greatest stem cell battle of all time. The campaign alone will cost around $50 million, more than many government programs in their entirety.

A big turnout is our only chance. We know that progressives among all parties support stem cell research, and the more information they have about our program, the more they tend to support it. If we have lots of progressives voting, and clear information available, we can win.

But progressive turnout drops as much as 20% in a non-Presidential election.

So if Bob does decide to lead, (and really nobody else can) I hope he will try for a 2020 initiative. That Presidential election year is our best shot; also our last, because after that, the money for the original program will all be gone.

Would this be a wise investment?

Listen to personal messages from three winners of the Nobel Prize:

Paul Berg: "The creation of the California Institute for Regenerative Medicine (CIRM) (is) a bold initiative by the citizens of California ... an exceedingly promising scientific breakthrough for improving the health of Californians.

Paul Berg (med.stanford.edu).

"During its eleven years ... (CIRM-funded) scientists achieved world leadership in extending ... stem cell biology to exploring opportunities for human cures or treatments. Many clinical trials for treating a wide variety of human afflictions are in progress or imminent."

"Now is not the time to slack off support for bringing the advances of the past eleven years of basic and applied research of stem cells to clinical fulfillment."

"We are near the "top of the mountain," but we need continued support to reach the pinnacle — providing cures for the ills that still plague humankind."

David Baltimore: "CIRM money has allowed California to be in the vanguard of stem cell research. It has been money without artificial restrictions on its use, allowing each scientist to take his or her personal approach to the subject. It has funded not only the development of basic understanding but the more difficult and chancy work of clinical investigation of novel approaches to therapy."

"Compared to Federal support of work in this new and highly promising area of research, CIRM support has been ... wide-ranging — a model for the country."

Shinya Yamanaka: "CIRM is a global leader of stem cell-based basic research and clinical application, providing great hope to patients with intractable diseases. I believe CIRM's initiatives will deliver innovative therapeutic options to solve the patients' most pressing needs."

"The California stem cell program has been helpful to many scientists during a time when obtaining funds for research has been increasingly difficult. Without such funding, research cannot go forward."

"The contribution to science for which I received the Nobel Prize, induced Pluripotent Stem cells, would not have been possible without research funding, and the California Institute for Regenerative Medicine is a source of help for many."

"It also encourages international cooperation between scientists, allowing the sharing of valuable information. It is my hope that the people of California will choose to continue this useful institution."

Five billion dollars.

How do we gather that money for research? It will require three achievable steps, the first of which is already underway, to be ready in case we go for it.

We must remind people what CIRM has already done. That is the purpose of this book; that is why patient advocates have already begun to organize in their communities: and why Americans for Cures Foundation (a continuation of the first campaign) is already reaching out: holding meetings and making new friends.

Second: if a ballot measure is selected, we must gather signatures to place it on the ballot. As you recall, the first measure required 650,000 (based on population); but we gathered 1.1 million to send the message that we were in earnest.

Third: With the help of public information ads to reach the general public, and advocates contacting every group we know, we will provide a reasoned basis for the California voters to vote "YES!", and continue our quest for cures.

Can we do it? Absolutely! In 2004, 1.1 million voters signed petitions to put Prop 71 on the ballot, and 7 million (59.1% of the voters that year) voted YES![6]

I remember a wonderful song of 1940, written by Charles Tobias in response to the attack on Pearl Harbor. Referring to our previous victory in WW1, he said:

"We did it before, and we can do it again, yes, we can do it again — we did it before — and we'll do it again!"[7]

Want to help? Contact: http://americansforcures.org/[8]

Remember: California's stem cell program was achieved by patient advocates exactly like you: people with a caring heart, the willingness to work, and the feeling that obstacles are just exercise.

---

[6] https://www.cirm.ca.gov/sites/default/files/files/about_cirm/External_Review_OOC.pdf

[7] http://www.huffingtonpost.com/don-c-reed/might-mouse-and-medical_b_2835956.html

[8] http://americansforcures.org/

David Baltimore (en.wikipedia.org).

Shinya Yamanaka (personal photo).

Think of it as strenuous fun: a vigorous adventure. You can help as much or as little as you want. If you, like me, are the quiet sort, a lot of the work you can do on your own, like writing letters of support. Or you can provide the gift of your time, sharing the noise and companionship of like-minded energetic folks.

If you support research for cure, this is the hour; this is the opportunity. Want to honor a loved one's memory? What could be a better living legacy?

Mary Bass (personal).

Would you like to help advance women's involvement in science? Listen to Mary Bass, Executive Director of Americans for Cures Foundation:

"It was a delight (but not a surprise) to discover that there are so many women at the forefront of this medical revolution: Dr. Judy Shizuru and Dr. Maria Grazia Roncarolo at Stanford University, pursuing cures

Yimi Villa (personal).

for babies born with a fatal condition called "Bubble Boy Disease"; Dr. Catriona Jamieson at the University of California, San Diego, who is pioneering new treatments to fight leukemia; or Dr. Jane Lebkowski at Asterias Biotherapeutics, who is developing a breakthrough therapy for severe spinal cord injury — in some cases, giving teenage quadriplegics the ability to lift weights again. It is these visionary women scientists, with their passion, intelligence, experience and commitment, who will continue to lead us into the scientific frontiers that hold historic potential to improve the health and well-being of societies worldwide

"I love my work with the cornerstone clinical trials going on in California as a result of Prop 71. I wake up every morning feeling empowered, ready to help change the world, proud to be a part of the fight for stem cell therapies, to be a part of these incredible advances in science and medicine."

Want to help lower-income minorities change the world? Here is Yimy Villa, Program Coordinator, Americans for Cures Foundation:

"The California stem cell program has played a defining role in my life. Once I received my Bachelor in Science from the University of California at Irvine, CIRM funding (through its focus on recruiting students from underrepresented minority groups) gave me the opportunity to pursue a Master's degree in Science. The CIRM Bridges Program I participated in at San Francisco State University, similar to programs CIRM funds all across the state of California, played an integral role in forging my career in stem cell research advocacy. Thanks to these CIRM funded programs, California has a new generation of scientists, doctors, and stem cell research advocates focused on improving the human condition."

— Yimy Villa, Program Coordinator, Americans for Cures Foundation.

We should preserve, protect and enhance the California stem cell program: it is the pride of a state, the glory of a nation, and a friend to all the world.

59 FORTY-TWO CALIFORNIA CLINICAL TRIALS

And now for something really special. Here are the actual clinical trials, new stem cell therapies approved for human testing by the FDA. Gathering the related information was a huge chore, accomplished by Mary Bass. Not only did she reach out to all the various scientists involved, but she wrote brief "people talk" descriptions of the disease and other pertinent information. Updated: 6/29/2017

Her full write-up is much larger, and better organized. You can see it at:

https://americansforcures.org/prop-71-funded-clinical-trials/

For more information, contact: Mary Bass, Executive Director, Americans for Cures Foundation: mbass@americansforcures.org

Each entry below contains: Disease, Principal Investigator, and Institution. Duplications are different attempts by the same people on the same conditions.

1. Age-Related Macular Degeneration: Mark Humayun, David Hinton, Dennis Clegg; USC, UCSB
2. Amyotrophic Lateral Sclerosis (ALS); Clive Svendsen; Cedars-Sinai
3. Ataxia; Joel Gottesfeld; Scripps
4. Blindness/Vision Loss; Pete Coffey; UCSB
5. Cancer, Colon; Chris Takimoto; Forty Seven, Inc.*
6. Cancer, Solid Tumors; Irving Weissman, Branimir Sikic; Stanford**
7. Cancer, Solid Tumors; Dennis Slamon, Zev Wainberg; UCLA
8. Chronic Granulomatous Disease (x-CGD) (severe immune disease); Donald Kohn; UCLA

9. Heart Failure Rachel Smith, Michelle Kreke; Capricor, Inc.*
10. Heart Failure, Duchenne Muscular Dystrophy (DMD); Deborah Ascheim; Capricor, Inc.*
11. HIV; John Zaia, Amrita Krishnan; City of Hope, Sangamo Biosciences*
12. HIV; Geoff Symonds; Calimmune, Inc.*
13. HIV-related lymphoma; Joseph Anderson, Mehrdad Abedi; UC Davis
14. Huntington's Disease; Vicki Wheelock, Jan Nolta; UC Davis
15. Glioblastoma; Anthony Gringeri; ImmunoCellular Therapeutics*
16. Glioblastoma; Stephen Forman, Christine Brown; City of Hope
17. Kidney Disease/Failure; Jeffrey Lawson; Humacyte Inc.*
18. Kidney Failure/Transplant; Samuel Strober; Stanford
19. Leukemia; Dennis Carson, Catriona Jamieson, John Dick; UCSD, University of Toronto
20. Leukemia, Chronic Lymphocytic; Catriona Jamieson, Thomas Kipps; UCSD
21. Infertility/Metabolic Diseases; Renee Reijo Pera; Stanford
22. Metastatic Melanoma; Robert Dillman; Caladrius Biosciences, Inc.*
23. Metastatic Melanoma; Antoni Ribas; UCLA
24. Melanoma, Sarcoma; Antoni Ribas, Zoran Galic; UCLA
25. Myelofibrosis (blood disorder); Catriona Jamieson; UCSD
26. Myelofibrosis; Catriona Jamieson; UCSD
27. Myelofibrosis; Catriona Jamieson; UCSD
28. Osteonecrosis (bone disease); Nancy Lane, Wei Yao; UC Davis
29. Polycythemia, Vera and Essential Thrombocythemia; Catriona Jamieson; UCSD
30. Polycythemia, Essential, Thrombocythemia; Catriona Jamieson; UCSD
31. Pulmonary Hypertension; Michael Lewis; Capricor, Inc.* Cedars-Sinai
32. Retinitis pigmentosa; Henry Klassen; UC Irvine
33. Retinitis pigmentosa; Henry Klassen; jCyte, Inc.*
34. Severe Combined Immunodeficiency (SCID): ADA-SCID; Donald Kohn UCLA
35. SCID x-Linked; Judith Shizuru, Morton Cowan, Jennifer Puck; Stanford, UCSF
36. SCID x-Linked; Brian Sorrentino; St. Jude's Children's Research Hospital

37. Sickle Cell Disease (SCD); Donald Kohn; UCLA
38. Spinal Cord Injury (Cervical); Jane Lebkowski; Asterias Biotherapeutics*
39. Spinal Cord Injury (Cervical); Jane Lebkowski; Geron, Inc.***
40. Spinal Cord Injury; Martin Marsala, Joseph Ciacci;UCSD;(Funding from Neuralstem, Inc.*)
41. Type 1 Diabetes; Howard Foyt, James Shapiro; UCSD, University of Alberta, ViaCyte, Inc.*
42. Type 1 Diabetes (Recent Onset); Douglas Losordo; Caladrius Biosciences, Inc.*

---

* Private Companies
** Derived from initial Disease Team grant (PI: Irv Weissman, Stanford University)
*** IP rights purchased by Asterias Biotherapeutics, Inc.

60 GATHERING OF CHAMPIONS

Morning, June 15, 2017.

Standing on the glass-walled skyway between the Westin Hotel and the Boston Massachusetts Convention Center, I watched a new world wake up.

Below me stretched the 15th Annual Meeting of the International Society of Stem Cell Research (ISSCR).

Everywhere was something un-missable, another adventure waiting, led by people who believed chronic disease could be defeated, and the world made well.

The ISSCR conference was a gathering of champions.

Like East Coast superstar George Daley. The former President of the ISSCR was there to speak on blood stem cells and to be honored for his decades of scientific contribution. One invaluable gift from Dr. Daley is his ability to sum up the latest advance or controversy — whenever something major happened in the field, I would look up George Daley and see what he had to say about it. He would always help sweep away the mists, clarify the confusion. In 2010, for example, he was one of only five scientists chosen to testify on behalf of the entire field of embryonic stem cell research, when it was under attack in a national lawsuit.[1]

You can also know a person by the company they keep. Dr. Daley is the husband of Amy Claire Edmondson — formerly chief engineer of R. Buckminster Fuller, inventor of the geodesic dome, and a philosophy

---

[1] http://www.ascb.org/wpcontent/uploads/2015/10/George_Daley_testimony_before_Senate_Appropriations_Committee.pdf

of world unity. Her book, *A Fuller Explanation*, is an in-depth examination of the work of one of the most important thinkers in world history.[2]

Everywhere you looked was another stem cell star: like Sean Morrison, now of the University of Texas, but formerly from Michigan, where he had been one of the primary fighters for freedom of research during that state's epic (and successful) political battle: Proposal 2.[3]

Canada's John Dick, all white-haired dignity and stature — the way every scientist must wish they looked, and every white hair stood for an accomplishment — was being honored by the ISSCR Tobias Award. John Dick has been a central moving force in stem cell research (especially cancer work) for more than 30 years …

Rudi Jaenisch was there, he who had worked with Beatrice Mintz (now 96) on the first transgenic mice. They established that DNA could be altered, bringing the possibility that chronic diseases could be cured. He had helped the world understand the crucial difference between therapeutic cloning for cells (potentially useful) and reproductive cloning to make babies (dangerous for both mother and child, illegal from California to China) — we are lucky to have him.[4]

And Leonard Zon, whose lab in 2005 created the first animal model of a melanoma cancer — in a zebrafish![5]

My feet were killing me. I have an age-related condition, peripheral neuropathy, (which means your feet hurt), and the last two days had been one big walk-a-thon.

So many people I did not know but wished I did, champions in their own countries, like Cedric Blanpain of the Universite Libre de Bruxells, Belgium, here to talk about the "Cancer Cell of Origin and Tumor Heterogeneity."

And friends not yet met, like Singapore's Huck-Hui Ng — I had interviewed him at long distance for my first book on stem cells, and

[2] https://www.amazon.com/Fuller-Explanation-Synergetic-Geometry-Buck minster/dp/061518314X/ref=asap_bc?ie=UTF8
[3] https://www.ibiology.org/ibiomagazine/issue-2/sean-morrison-stem-cell-politics.html
[4] https://en.wikipedia.org/wiki/Rudolf_Jaenisch
[5] https://zon.tchlab.org/

here he was, talking about "Modeling Disease Using Human Organoid Systems."

The most moving speaker was Sanford Greenberg of the Johns Hopkins Wilmer Eye Institute. Blind since 19, he was a scientist/inventor challenging the condition — with a $3 million dollar golden prize (literally, $3 million in gold!) his gift to whoever discovered the cure for blindness by the year 2020.

The most interesting title for a talk? *Stressed Out: A Novel Approach to Cancer Immunotherapy*, by Laurie Glimcher, of the Dana Farber Cancer Institute.

My favorite "told you so!" moment? Dan Kaufman's talk on cancer and how to fight it: "Advances in Natural Killer Cell Development from Human Pluripotent Stem Cells"! Kaufman's efforts had been remarkable for a long time, and I was glad to see him in the spotlight.

Most useful technique for networking? The "Meet-up Hub," where you could meet (for example) the German Stem Cell Network, all in one place, for 45 minutes.

How do you make a target of ALS? Justin Uchida of USC was using skin samples from patients as models of Lou Gehrig's disease.

Perhaps the most important company to help a scientist get a job? **Job Match**. I do not know them, and cannot vouch for them. But a job-hunting agency specializing in scientists? That is a brilliant concept, and very needed.[6]

Want to know how to work with the FDA? Ask Mercedes Serabian, of the U.S. Food and Drug Administration.

Parkinson's disease? Four experts, from different corners of the world:

a. Claire Henchcliffe of Cornell Parkinson's Institute, speaking on "Critical Needs for Optimal Treatment of PD";
b. Roger Barker, University of Cambridge: "the Use of Fetal Dopamine cells to treat Parkinson's";
c. Viviane Tabar, of Sloane Kettering, "Embryonic Stem Cell-Derived Dopamine Neurons for Treatment of Parkinson's";
d. Brian Fisk, from the Michael J. Fox Foundation, "Development of

---

[6]https://jobs.newscientist.com/

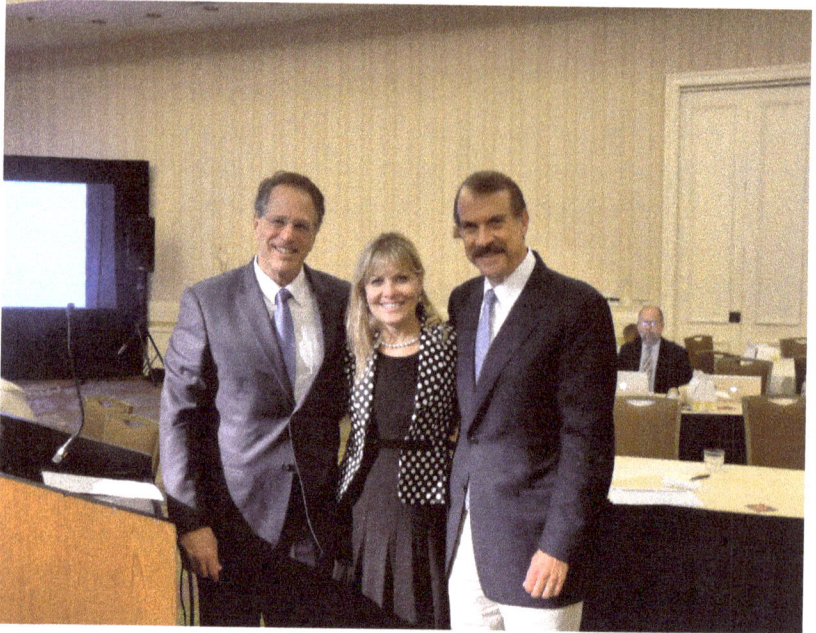

Bob Klein, Danielle Guttman-Klein, Jonathan Thomas (Californiastemcellreport.blog.spot).

Cell-based Therapies" for PD.

Looking down through the glass wall, I saw a huge mechanical hoist lifting a basket, inside of which was a young man frantically waving his arms.

Signs and display booths filled the cavernous hall; hundreds of companies vying to serve an emerging field.

In the distance, hundreds of white poster boards: what a joy it had been to walk around and listen to the young scientists as they stood by their posters.

Canada's Mina Ogawa did a "poster teaser" (30 seconds to explain her approach) presentation on "Modeling Cystic Fibrosis for Drug Profiling on Hepatic-Biliary Organoids from Human Pluripotent Stem Cells" — that title shows why it is such a good idea for scientists to learn to "talk short!"

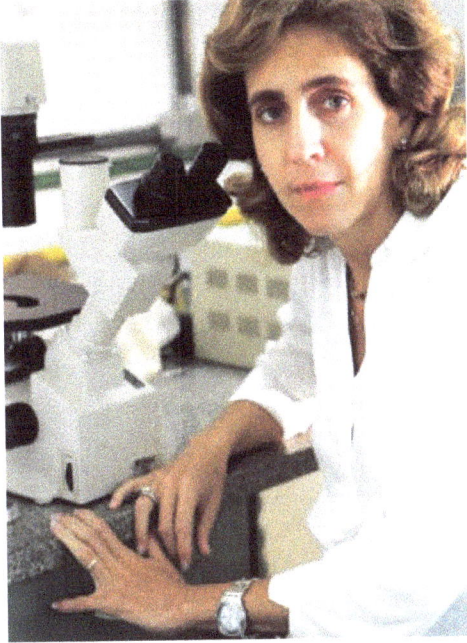

Lygia Perreira of Brazil (LP personal photo).

Canada, BTW, is packed with top-notch researchers: like Connie Eaves, battling the negative stem cells of breast cancer in Vancouver, British Columbia.[7]

And to the South? Lygia Perreira was there, one of the champion researchers of Brazil, but I missed connecting with her ...

Some of the presenters talked too long, running overtime, heedless of the moderator's increasingly frantic hints. Some did not know how to make a wrap-up; instead, even as they ran out of time, they said exactly what they had planned to say, only faster and faster, so the words spewed out in a verbal blur ...

Still it was better to have too much progress to report than not enough.

---

[7] http://www.bccrc.ca/dept/terry-fox-laboratory/faculty/dr.-connie-eaves

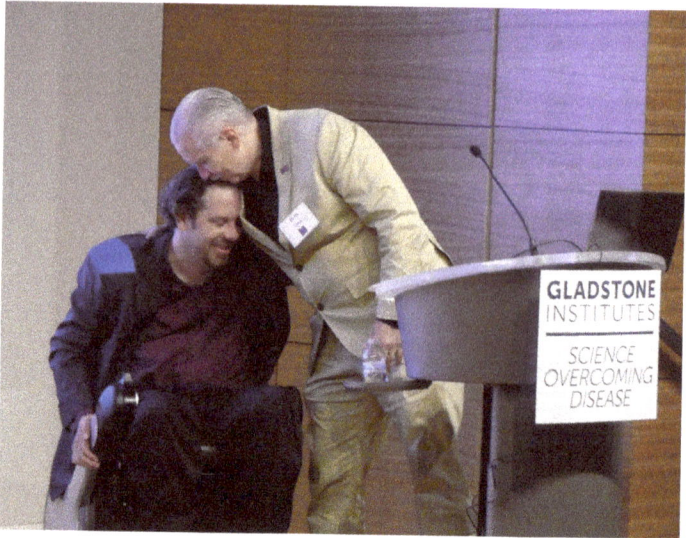

Don and Roman Reed (personal).

Everything was changing, in a field so young that folks were still alive who had been with it from the beginning.

Like Elaine Fuchs, recipient of the McEwen Award, honored by the ISSCR for her 40 years of research with skin stem cells, which she summed up in 20 minutes.

Dr. Hans Clever, the next President of the ISSCR, used cartoons to illustrate a very listenable lecture on organoids. Instead of just saying that the 3-dimensional bits of living glop let us duplicate the growth of the inside of an intestine, the cartoons showed us how it worked, in a non-gross way.

I remembered Bob Klein saying "a good visual lends wings to understanding — particularly in these non-verbal times."

Being able to stay only two days of the five, I missed more than half of the show.

I heard Nobelist Shinya Yamanaka's talk on ten years with his invention, Induced Pluripotent Stem Cells, and loved how the Japanese government was backing up his research so strongly.

Regretfully I missed Doug Melton's presentation on using embryonic stem cells to fight diabetes.[8] In California, we naturally hear about a lot about ViaCytes' work in this area, because they are our neighbors, and are funded by CIRM. But Melton's efforts are said to be terrific, and I know I join the community in wishing him all success.

Aging issues? China's Weiqi Zhang was going to speak on "Investigating Premature Aging Using Stem Cell Models." USC's Albina Ibrayeva spoke on "Origins of Age-Related Neurogenesis Decline."

The use of one therapy to treat many? Sweden's Ollie Lindvall was going to present on "Replacing Neurons by Transplantation in the Diseased Human Brain," potentially important to many nerve disorders.

Jane Lebkowski! Everybody knows Dr. Jane, one of the pillars in paralysis research since the Geron days, the world's first attempt (ongoing) at embryonic stem cell therapy, research initially funded by "Roman's law."

Out of ten questions in Q and A, nine were for Dr. Jane. She might walk with a cane, but her mind was young and agile; she gave speeches on two different subjects: one on paralysis, the other on the eye.

One worry about paralysis research is the difference between new (acute) injuries and older (chronic) ones. Scientists like to work with new injuries; the chances of improvement are greater. Also, the spinal cord scar is a barrier to new cells. Across the world there were millions of people with old injuries — did having that scar condemn them to a lifetime of paralysis?

I asked if it was possible to **remove the scar** after an injury, turning what had been a chronic injury back into a new one, perhaps more suitable for stem cell therapy? She said it was a good question, and people were working on it.

I headed toward the lobby to check out.

My most fun memory? I had lunch with advocates like Adrienne Shapiro (tireless worker for sickle cell anemia relief), Mary Bass, Yimy Villa, Bob Klein, and Larry Soler. Larry is famous for his work to support diabetes cure research. It was Larry who worked side by side with Bob to raise $1.5 billion for diabetes research in 2002. Irv Weissman ate with

[8]https://hsci.harvard.edu/news/stem-cells-billions-human-insulin-producing-cells

us, Stanford's cheerful king of stem cells, and more whose names escape my limited recall …

Speaking of Bob, it was fun to watch him try to move through the crowds here; everybody recognized him; he was mobbed like a rock star.

And of course, everyone wanted to know: **would there be a Prop 71, Part Two?**

There were no outright answers; no decisions had been made. But here was Bob, having as many meetings with stem cell leaders as he could cram into a day …

So with all these champions sitting around the table, what do I remember most vividly? What word of wisdom, or intellectual problem most engaged me?

Well, actually, it was the food.

There was something called a "slider," like a miniature hamburger but with a thick square of high quality meat and cheese stacked between the buns. The whole thing was silver-dollar-sized across, but at least six inches thick, a leaning tower of food, seemingly intended to be shared.

The problem was, I had no idea how to divide it.

So, I waited, figuring somebody younger would negotiate the awkwardness.

Mary Bass was sitting on my right. The slider was between us.

I waited.

So did Mary.

At last she accepted the challenge, picked up her knife and fork, and attempted the division of the food which lived up to its name.

As though trying to escape, the slider slid across the plate.

But Mary persevered, and in the end, there was one nice neat half of a slider with the other half reduced to a little forest of meat slivers and squashed bread.

I looked at Mary.

"I was waiting for you to do that," I said.

"I knew you were!" she said, starting to laugh.

"And see how well it turned out?" I said, and stabbed the neat half of the slider for myself.

I wished I could have thanked Sally Temple, this year's President of ISSCR, and Nancy Witty, Chief Executive Officer, and Kaye Meier,

Director of Policy, and all their co-workers, supporters and friends: the ISSCR Conference is a necessary floodlight on vital research, and this years' was especially excellent.

I scrambled aboard the airplane, suddenly eager to get home.

Gloria would be waiting at the airport, and I wanted so much to see my son and grandchildren. Katherine had just had a Brazilian Jiu Jitsu match (she won, pinning the unfortunate young man in one move, even though he pulled her hair), Jason had a champion baseball game coming up; Roman Jr. had broken his ankle (twice!) playing basketball, Jackson and his mother Desiree were relocating to Las Vegas, the University of Nevada at Las Vegas (UNLV) where she had just been named the nation's one and only female Hispanic Athletic Director ...

On the flight home I studied my near-incomprehensible handwritten notes and the various ISSCR handouts for five hours, trying to understand what I had learned.

And then, when I could not stand being virtuous for one more second, I turned on the free section of the airplane's entertainment.

They were featuring three episodes of *The Big Bang Theory*.

I watched them all.

61 GOODBYE, HELLO!

June 19, 2017

For two, the day would be an ending; for one, a new and tremendous challenge.

It was the Independent Citizens' Oversight Committee meeting. There were numerous outstanding grants to be funded, (and they were) but this meeting was about change.

First came an emotional blow: the official retirement of James Harrison, long-time chief lawyer for the California Institute for Regenerative Medicine.

When I heard James was leaving, my first thought was: "Don't go!"

I knew, logically, that James was an adventurer, and that for him, in Tennyson's words, "there were yet great deeds to do." His organization, Remcho, Johansen and Purcell[1], was massively involved in democracy. Some of their clients included the California Senate and Assembly, the California Teachers Association, Consumer Attorneys of California and many more.

And, he would be replaced by CIRM attorney Scott Tocher, respected by all.

But even so, to lose James Harrison?

He had been with us, literally, since before CIRM began. When Bob Klein wrote Prop 71, shaping an incredible dream into words, James Harrison was at his side.

California knew James Harrison. He had worked on six important initiatives, including TV star Rob Reiner's Proposition 10, the California

---

[1] http://www.rjp.com/index.cfm/about/overview/

Children and Families Initiative, which turned the tobacco settlement into good use. When Prop 10 was successful on the ballot, it was immediately attacked as unconstitutional. But James was there, and California won. James had been involved in so many pro-people efforts and initiatives; I hoped he would write a book about them.

But above all was the California stem cell program. At every step of the way, James Harrison was in our corner. When the anti-research opposition sued to shut us down, James was there, fighting back, successfully. One small but typical example: when I was writing about the CIRM, and I had a legal question, I would call up and leave a message for James, and he would unfailingly get back to me. This was nothing he was paid for;he just cared about the cause.

And now I had to stand up in the meeting, and say goodbye. I could feel my voice trembling, and hurried to get through:

"But whatever sadness I feel right now, I take comfort in one thing. I know that when and if there is a part two for Prop 71, James will be there, fighting beside us once again — we will give him no choice!" So —

James Harrison (Cirm.ca.gov).

"Vaya con Dios, James Harrison. Go with God; our hearts are with you, always."

We were also losing our current President, Randy Mills. He had come to us as a biomed business expert, the CEO/President of Osiris. His company developed one of the first stem cell therapies, Prochymal, to fight graft-versus-host disease, saving children's lives. With his business/medical background, he looked at CIRM's unique and wonderful organization, and thought: this could be better.

"I did not like you very much at first," I said, "You came to what I consider the greatest government agency of all time — and you wanted change?"

There was about thirty seconds of laughter, and Randy's comeback, "Well, I liked *you* from the start!" But, I was not altogether joking. Change is never easy.

All the other Presidents, like Zach Hall and Alan Trounson, had focused on the science, but it seemed like Randy Mills wanted to change everything.

But, he brought improvement. For scientists wanting grants, Dr. Mills' changes *made the process easier.* Before, a scientist might try for a grant at a time that was not optimal. The scientist might not be completely

Randy Mills (Cirm.ca.gov).

Maria Millan (Cirm.ca.gov).

ready; the CIRM might not need his/her particular expertise just then. Now, the application schedule was on a rolling basis. If one opportunity for funding was missed, another was near.

How important?

Randy's changes brought cure closer, sooner. According to no less than Jonathan Thomas, chair of the ICOC, it meant "an 82% reduction in approval time for clinical trials, a 3-fold increase in the number of clinical trials, and a 65% reduction in the time it takes to enroll (patients for) those trials."

Now, Randy Mills was returning to the private sector, to become President/CEO of "Be The Match Biotherapeutics," an organization matching cord blood donors with recipients. He would bring change — and improvement — wherever he goes.

And who would be the new President? For me, the choice was clear.

Dr. Maria Millan was currently vice-President for CIRM's therapeutics. If the vote went her way today, she would fill in as interim (temporary) President. Did that mean she was a shoo-in for the Presidency? No. That decision would come later.

But one step at a time: today, she had to be elected to the interim slot. Her strengths? Communication, compassion, and overwhelming competence.

Communication? If Maria Millan used a big word, it would be made clear. If she said, "haematopoietic disorders," it would be followed by: "you know, things that go wrong with the blood." I never felt talked down to; I just understood her. Such clarity is vital; if we in the public are to support something, we must understand it.

Compassion? On her desk was a small box, containing 500 hospital wristbands. Each one was a person she had operated on. If a scalpel was a sword, she had gone with those people into the valley of the shadow of death, and had fought for them.

Competence? Dr. Millan had trained in general surgery and transplant immunology at the Beth Israel Deaconess Medical School at Harvard: she became Associate Professor of Surgery (and Director of the Pediatric Organ Transplant Program) at Stanford. As Chief Medical Officer for Stem Cells Inc., she oversaw the launch of the company's first clinical trials, and was head of their liver program.

A part of CIRM since 2012, she helped build some of its most powerful components, like the Alpha Stem Cell Network. There would be no on-the-job training required; she was ready to go.

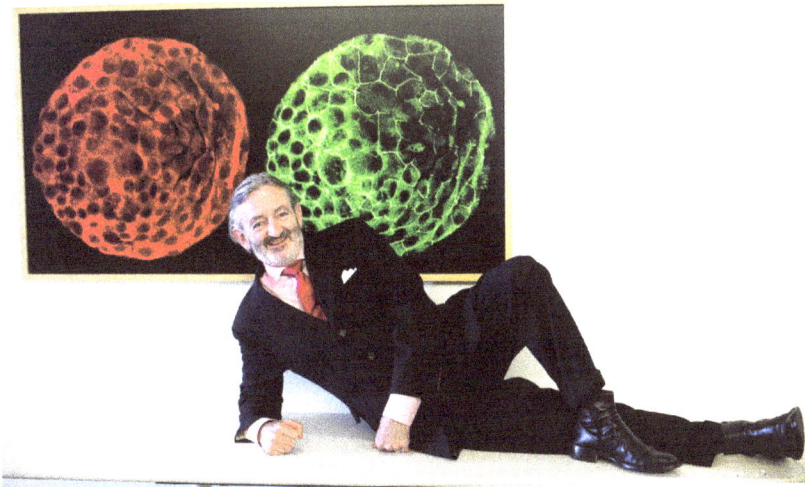

Kevin McCormack (CIRM photo).

Should Maria Millan be selected as temporary President for CIRM?

"We should keep her forever," I said — and there was no opposition.

One by one the vote was called. It was unanimous in support of Dr. Millan.

CIRM director of Communications Kevin McCormack spoke for all:

"May I be the first to say, "Welcome, Madame President, and I have been wanting to say that for nearly a year!"

62 THE ANSWER

For the past several years, there has been one burning question in my mind: will there be a Prop 71, Part Two: will Bob Klein lead another effort to fund stem cell research and regenerative medicine, this time to the tune of $5 billion?

With the publication schedule tightening for this book, I needed to have an answer for the reader. Not easy! Though Bob's office is just 50 feet down the hallway, his usual state of affairs is best summed up by the word "swamped." Accordingly, I put in a formal request to see him, via Elizabeth Taffeen, his right-hand person.

The next day, there was a momentary lull in the whirlwind. I walked past the real estate offices of Chad, Maati, Mimi, Alan and Linda, friends at Klein Financial who make Bob's stem cell efforts possible, and sat down across from him, at the black-topped mahogany desk where Proposition 71 began, so many years ago.

I asked the question. And here, as best I can decipher my scribbles, is the answer.

"Technically, our program began in December, 2004. But since ideological lawsuits held us up for nearly three years, we did not begin funding grants until the Summer of 2007. Accordingly, if the public is polled in winter of 2019, they will have had 12 years to review CIRM's performance."

"The people of this state will be asked to judge if the California Institute for Regenerative Medicine has indeed made a historic breakthrough in medicine. I believe we have had a revolutionary impact on human suffering, attempting to protect every family from the devastation of chronic disease. In FDA-approved human trials, we have a proven record of our battle to lift the human condition."

"Scientists, doctors and patient advocacy groups will bear witness to the improved standard of living regenerative medicine can bring, to every American and citizen of the world. The California public will, I pray, understand our state's unique role in opening a path for every family to a new future of health and quality of life."

"And so the decision is in the hands of the California public. They will decide, when polled in the Winter of 2019. If they strongly support the continuing efforts of the CIRM — *then I am fully committed to a 2020 initiative* to renew the funding of the California Institute for Regenerative Medicine."

Robert N. Klein, July 11, 2017

63 A NOBEL PRIZE FOR BOB KLEIN?

As a non-scientist, I have no stature to recommend Bob Klein for the Nobel Peace Prize, Earth's highest honor. But if I could, I would, and here is what I would say.

Bob Klein has changed the world for the better in three major ways, including:

1. Increasing the availability of affordable housing for people in need;
2. Contributing integrally to the raising of $1.5 billion in diabetes research;
3. Raising $3 billion for stem cell research, (and $3 billion interest on the debt) by revolutionizing the funding mechanism of medical inquiry.

Bob is the President of Klein Financial Corporation, a real estate investment banking consulting company focused on affordable housing finance and development. An expert in the field of affordable housing, Bob conceptualized, developed and managed legislation that created the California Housing Finance Agency (CHFA), providing low-cost loans to lower-income families, and making affordable apartments available. He served six years on the CHFA governing board. When he left the board, the CHFA had approximately $8 billion in financing outstanding, an S&P bond rating of AA, and had earned numerous national awards. The agency continued at the pace of approximately $1.5–$2 billion per year, including matching funds from governmental programs. Klein Financial has never accepted a loan from the CHFA, the agency Bob began.

Bob Klein and family.

In 2002, as the primary U.S. Senate/House negotiator for JDRF, Klein marshaled unanimous consent for a $1.5 billion appropriation for diabetes research and treatment: the first-ever stand-alone healthcare bill to pass both the U.S. Senate and House unanimously. Because the funding also maintained diabetes clinics on Native American reservations, their only treatment source, lives were saved.

But it was his work to design and implement the California stem cell program that most influenced our planet. In 2003, Klein authored and chaired the campaign for Proposition 71: the California Stem Cell Research and Cures Initiative, personally donating more than three million dollars to the campaign, and helping raise much more. In 2004, Proposition 71 passed with a 59.1% majority — establishing a new state agency, the California Institute for Regenerative Medicine (CIRM).

Serving without a salary for the first six years, Bob chaired the board of directors: the Independent Citizens Oversight Committee. His chairmanship led to the creation of 12 Research Institutes and Centers of

Excellence in California. His requirement of matching grants for the new facilities leveraged California's investment of $272 million into a $1.15 billion monetary return for the State.

With the stem cell program written into the California Constitution, a safe haven for advanced research was established.

CIRM has funded research leading to at least 42 FDA-approved human clinical trials in progress or pending. Therapy targets include paralysis, diabetes, blindness, Huntington's, heart failure, HIV/AIDS and more. CIRM-funded scientists have published 2,500 peer-reviewed "medical discoveries," in documented literature.

Challenges were many. When Sacramento proposed major overhauls to the governance structure California voters had approved, Klein negotiated a consensus agreement with the legislative leadership, incorporating additional ethics and procedural safeguards into the agency's operating bylaws. Standards and peer review processes were established in collaboration with the National Academy of Science. For example, intellectual property agreements with CIRM required an Access Plan for low-income citizens, and public clinics and hospitals in California were assured discounts when therapies would become commercially available.

Klein's belief in international research cooperation led to Partnering Agreements with 14 foreign governments, from Japan to Australia, as well as agreements with U.S. states, institutes and foundations, including the NIH Research Hospital.

The CIRM is a paradigm change in funding medical science. Its long-term bonds raise a critical mass of funding that drives breakthrough research through the development stages and into the clinic, while spreading the costs over decades, paid for by citizens who will benefit. This medical financing model encouraged other states and countries to compete, driving them to develop funding of at least another $5.6 billion in stem cell research and therapy development beyond CIRM's funding and matching funds — just during Klein's term as Chairman.

Standards and policies were designed and implemented in the full light of public view, necessitating a staggering number of multi-hour meetings. As transcripts will attest, it was hard to find any meeting without Bob Klein in attendance.

No one can predict the pace of science; no one can say when cures will come. But change for the best is already happening. For example:

Researchers at UCLA performed what may well be the World's first stem cell cure: 18 small children cured of Severe Combined Immunodeficiency, "Bubble Baby" disease. This therapy development was performed at the Eli & Edythe Broad Center for Regenerative Medicine — an institute created with CIRM funding.

Such success brings closer the ultimate goal of medicine: to make people not just better, but well: to heal the patient, not merely maintain him/her in their suffering.

The future of regenerative medicine has been strengthened by the Bridges Program, which empowered over 700 of California's brightest lower-income students at city and state colleges, bringing them into the nation's leading labs.

In 2010, an External Advisory Panel (EAP) of eight world-renowned experts concluded that: "In a short few years, CIRM has created a robust, world class stem cell research effort in California, with a greatly expanded workforce, state of the art facilities and the requisite physical and intellectual infrastructure needed to accomplish its scientific goals."[1]

On June 23, 2011, Bob Klein retired from the Chairmanship of the Independent Citizen's Oversight Committee of CIRM, saying:

"It has been the privilege of my life working with this board and the staff of CIRM. We are on a mission for all our families, in our state, our country, and the world. With the tremendous outpouring of dedication and effort of all involved, including especially the patient advocate groups, I believe we will be successful ..."

The first action of his replacement Jonathan Thomas was an ICOC vote that Bob Klein would receive the title, "Chair Emeritus," to honor his contribution.

But that was by no means the conclusion of Bob Klein's public service.

After two full terms on the Board of Directors of the International Juvenile Diabetes Research Foundation, Klein now serves on the Board of Chancellors of that organization. He recently retired from the board of Genome Canada, a Canadian government research agency with $1.4 billion in authorized research. He was co-chair of the Stem Cell and Regenerative Medicine working group of the Canada/California Strategic Innovation Partnership, which initiated the Cancer Stem Cell Consortium, a strategic partner of the California stem cell program.

Bob previously served on the Board of Global Security Institute, dedicated to reducing the global risks from nuclear weapons. He has served as Vice-president of the Board of Directors of the State of the World Forum, an international organization focused on global economic development, and helped develop California's National Historic Site Restoration Program.

Bob has a Bachelor of Arts in History with Honors from Stanford University and a Juris Doctorate degree from Stanford Law School. He is a member of both the California Bar Association and the American Bar Association. He retired as a lieutenant from the United States Air Force.

Bob lives in Northern California with his wife, Danielle Guttman-Klein, and has four children: Lauren, Robert, Jordan (recently deceased due to complications related to Type 1 diabetes) and Alyssa.

The world has recognized Klein's accomplishments, including:

Time Magazine included Bob in the "World's 100 most influential people of the year" for 2005, also calling him "one of the 10 most influential thinkers of 2005";

Scientific American named Bob to their list: "The Scientific American 50";

A proclamation from the California State Legislature honored "his ... role in the successful negotiation of $1.5 billion of National Institutes of Health ... research funding for ... diabetes; his work on $3 billion in stem cell research funding and Chairing the California Institute for Regenerative Medicine (CIRM) ...";

Bob received Research! America's Gordon and Llura Gund Leadership Award in March 2010;

In 2010 he also received the Biotechnology Industry Organization (BIO)'s "International Biotech Humanitarian Award";

In 2011, Bob was honored by the International Society for Stem Cell Research's inaugural "ISSCR Public Service Award."

For his tireless efforts to ease suffering and save lives, Robert N. Klein is a deserving candidate for the Nobel Prize.

AFTERWORD: FOR MORE INFORMATION

David Jensen (Californiastemcellreport.blog.spot).

1. A must-read source is the **California Stem Cell Report by David Jensen**. Although he basically opposes the program, Jensen's coverage is exhaustive and honest; he tells both sides. http://californiastemcellreport.blogspot.com/

Paul Knoepfler (https://ipscell.com/).

2. **Paul Knoepfler** was America's first scientist -blogger — his material is always cutting-edge big issues. A real friend to stem cell research, Paul gives an annual $1,000 award to the most helpful person to the field. It is a good thing he is not eligible for his own award, or he would just keep winning it! Check out his column, "The Niche," at https://ipscell.com/

3. **ScienceDaily.com** is a treasure. It provides "people-talk" explanations of new experiments, and not just on stem cells. Suppose someone you knew had a chronic disease and you wanted to find out the top 30 or 40 experiments aiming at cure for that condition? Just visit the website, type the condition into the searchbox, and there it is! https://www.sciencedaily.com/

4. **Americans for Cures Foundation.** Google us, and take a look. If Bob Klein decides to go for another $5 billion in more California stem cell research, here is where you will find the most up-to-date info on that great battle. https://americansforcures.org/

5. **www.stemcellbattles.net** is my personal website. It has access to my continuing series of free short articles on the fight for funds and freedom.

You can reach me personally at: diverdonreed@google.com

Be sure to read my other book, "*Stem Cell Battles: Proposition 71 and Beyond: How Ordinary People Can Fight Back Against the Crushing Burden of Chronic Disease.*" You can get it cheap at Amazon.

Always remember: if we never give up, we can only win, or die. And everybody dies — so why not try?

All best to you and yours.

Don C. Reed

P.S. My cancer tests came back negative.

June 21, 2017

PERSONAL MESSAGE

Dear Friend Who Reads This Now:

Do you support stem cell research? If so, consider helping me spread the word. Please recommend it to friends and social media contacts. And if you would *really* like to help share the message...

Here are three important actions you could take.

1. Go to my website (www.stemcellbattles.net) and sign up to receive my free newsletter. That way, when I write something for Huffington Post, or my own weblog, you will automatically receive a copy.
2. Go to Amazon, look up this book, scroll down to the bottom of the page, and write a review. Nothing fancy is needed; a couple of sentences is fine. If you thought it was good, give it five stars. The more five-star reviews a book gets, the better placement Amazon will give it.
3. Have you read my previous book, *Stem Cell Battles: Proposition 71 and Beyond: How Ordinary People Can Fight Back Against the Crushing Burden of Chronic Disease*? It is very much like this one, only bigger and with more stories. It contains the first half of the struggle to pass and protect the California stem cell program, international efforts, and a bunch more. If you have a stem cell scientist friend or two, they are probably in this book, and might appreciate a copy.

If you need to contact me, my email is: diverdonreed@gmail.com.

Always remember: you are more important than you know. You make a difference; you matter. So plan your life and work the plan.

I wish you joy.

All Best,

Don C. Reed

THE END (Unless California votes for a Proposition 71, Part 2 — for $5 billion!)

P.S. For the ultimate resource on the California stem cell program, go to its website: www.cirm.ca.gov

INDEX

Aboody, Karen, 150–153
Aging, 93
Aivita Biomedicine, 170
Alligators, relocating of, 155
Alpha Stem Cells Clinical Network, 83–85
Alvarez-Buylla, Arturo, 264–266
Alzheimer's, 97
Americans for Cures Foundation, 332
American Heart Association, 61
American support for medical research, 196
Ants, 107–110
Aortic aneurism, 261–262
Arc, Joan of, 267–271
Aron, Gloster, 145
Arthritis, 118–120, 225–228
Asterias Biotherapeutics, 170
Asthma-bronchitis, 9
Autism, 127–129

Baccheta, Rosa, 121–122
Badie, Benham, 150, 198
Bae, Hyun, 220–224
Baldness, 29–32
Bales, David, 142
Baltimore, David, 297, 299

Bass, Mary, 193, 300–301, 314
Bauer, Gerhard, 73
Berg. Paul, 296–297
Big Bang Theory (TV show), 131–134 (Stars and Writers)
Biden, Joe, 149–153
Biden, Beau, 149–153
Biomed as career, 43–47
Blanda, Theresa, 249–251
Blau, Helen, 94–96
Blindness, 7, 33–37
Blurton-Jones, Matthew, 74
Bond, James, (Movie), 179
Boisen, Kris, 190
Brain tumors, 149–153
Brainstorm Cell Therapeutics, 247
Broad, Eli and Edythe, 10
Brown, Christine, 150
Brown, Rayilyn Lee, 269
Buck, Frank, 69–70

California Institute for Regenerative Medicine (CIRM), 44
California Stem Cell and Biotechnology Workforce Development Act, 45
Cancer, 53–58

Capra, Frank, 65
Chamberlain, Stormy, 147
Charo, Alta, 113
Chieh, Tsao, 157
Chen, Bertha, 259–262
Cheng, Alan, 20–21
China, 275–280, standards for trials, 278–279
Chronic disease, costs of, 294–296
Chuong, Cheng Ming, 31
City of Hope, 193–202
Clark, Dwight, 245–246
Clinical Trials.gov, 81, 303–305
Collins, Francis, 104–105
Connecticut, stem cell program, 141, 145–148
Cox, Doug, 176
Couture, Larry, 150
Cowan, Chad, 79
Critical Limb Ischemia, 13–16
Currais, Antonio, 100–102

Daley, George, 307
Darzacq, Xavier, 232–234
Deafness healing research, 20–21
Dealy, Carolyn, 147
Diabetes, 7, 87–92, 175
Dick, John, 82, 308
Dilgen, John Hudson, 23–24
D'Lima, Darryl, 119–120
Disability, voting rights, 235–237
Dolphin, 17–18
Doonesbury, 253–255
Doudna, Jennifer, 230–231
Douglas, Kirk, 96
Dracula (John Carradine), 225–228

Duliege, Anne-Marie, 287–292, with daughters, 289

Embryonic stem cell research, 60–63, 174–177
Epidermolysis bullosa, 23–27
Epilepsy, 145, 263–266
Evseenko, Dennis, 226–228

Fei, Li, 276, 279
Fetal cell research, 111–113
Fisher, Dan, 173–174
Food and Drug Administration (FDA), 84, 163–164
Fox, Michael J., 269
Funicello, Annette, 209–211
Furlong, Pat, 197

Gallegos, Michael, 173
Gardner, Mimi, 90
Gazit, Dan, 220–224
Gehrig, Lou, 245
George, Jacob and Davis, 167
Goh, Keng Swee, 156–157
Gold, Joe, 194
Gomperts, Brigitte, 10–12
Grabel, Laura, 145
Grant, Ulysses S., 96
Greenberg, Sanford, 309

Harris, Andrew, 103–104
Harrison, James, 117–118, 317–319
Harvard Stem Cell Center, 79
Hearing loss, 18–21
Heart disease, 49–52, 61–63
He Xu, Ren, 142
Heston, Charlton, 107

HIV/AIDS, 253–255
Hollenbring, Wilger, 68
Holmes, Oliver Wendell, 93
Humayun, Mark, 34–37
Huntington's disease, 70–74

Idiopathic pulmonary fibrosis, 10–11
Induced pluripotent stem cells, 26
Independent Citizens Oversight
    Committee (ICOC), 58, 115–120,
    317–322
International Society of Stem Cell
    Researchers (ISSCR), 81, 307–315
Investigational new drug
    application, 84
Invisible Man, (Movie), 131
IPEX disease, 121

Jaenisch, Rudi, 308
Jamieson, Catriona, 81–85, 250–251
Jenkins, Julia, 25
Jensen, David, 331
Jing, Naihe, 275–276
Jolson, Al, 5–7
Jovin, Thomas, 108

Kajimura, Shingo, 77–79
Kaufman, Dan, 56–58, 309
Keirstead, Hans, 169–171
Keller, Helen, 19
Kipps, Tom, 82
Klassen, Henry, 34–37, 193–194
Klein, Bob, 7, 87–88, 98, 199–202,
    294, 310, 323–324, 325–329
Klein, Danielle Guttman, 310
Kohn, Donald, 125–126, 243, 257–258
Koko (Gorilla), 259
Knoepfler, Paul, 332

Kozlovich, Cory, 198
Krause, Diane, 147
Kuo, Caroline, 257–258

Laird, John, 14
Laikin, Paul, 92
Lajara, Rich, 165–166
Lane, Alfred, 25
Lane, Thomas, 211
Langston, William, 108
Lebkowski, Jane, 301
Leukemia, 81–85
Li Ka Shing, 229
Lim, Chuan Poh, 158–159
Lin, Haifan, 147
Liu, Edison, 158
Lincoln, Abraham, 157
Liver, 65–68
Lomax, Geoff, 195–196
Loring, Jeanne, 211–212
Love, Ted, 240, 242
Lubin, Bert, 242
Lung disease, 10–12
Lupian, Vanessa, 67–68

MacDonald, Kristin, 196
Macular degeneration, 34–37, 175
Magnificent Seven, (Movie), 139
Malin, Jennifer, 195
Maple syrup urine disease, 67
Marine World Africa USA, 17,
    183–186 (orcas and sharks)
Maryland, state stem cell research
    program, 143
May, Michael, 165
McCormack, Kevin, 126, 321–322
Melanoma, 179–181
Melton, Doug, 313

Measles, 129
Mesenchymal stem cells, 15–16
Millan, Maria, 320–322
Mills, Randy, 83, 163, 193–194,
  319–320
Mini-brains, 127–129
Minnesota, regenerative medicine
  program, 141
Mintz, Beatrice, 308
Mirels, Lily, 230–231
Monuki, Ed, 73
Multiple sclerosis, 209–203
Muotri, Alysson, 127–129
Murphy, Chris, 145
Murry, Charles, 143
Muscular dystrophy, 197–198
Myamoto, Musashi, 135–137

Naegele, Janice, 145
Neurological diseases, 184–186
New York stem cell programs, 141,
  203–208, Times Square sign,
  208
Ng, Huck-Hui, 308
Nolta, Jan 15–16

Obama, President Barack, 59
Obamacare, 241
Obesity, 75–79
Ogawa, Mina, 310
Oklahoma, 173
Oro, Anthony, 25–26

Paddila Vaccaro, Evangelina, 1–3, 7,
  124–125
Paralysis, 7, 175
Parkinson's disease, 107–110, 231,
  269–271

Patterson, David, 205
Pence, Mike, 113, 215–217
Perreira, Lygia, 311
Personhood, 104, 174, 215–217
Plikus, Maksim, 29–32
Political obstacles to research, 59
Polycythemia Vera, 249–251
Portnow, Jana, 150–153
Pricing of CIRM-funded products,
  50–52, 194–195
Prieto, Francisco, 89
Proposition, 71, 7
Pryor, Richard, 209–210

Quintiles, 138
Quorum activity, 30

Reed, Charles, 93, 94
Reed, David, 94, 219–214
Reed, Desiree, 288, marriage
  proposal, 293–294
Reed, Don, 44, 162, 171
Reed, Gloria, 49–52, 97, 127, 225
Reed, Jackson, 315
Reed, Jason, 27
Reed, Katherine, 93, 171, 290, 315
Reed, Patty-81
Reed, Roman, 39–42, 187–191, with
  sons, 190, 312
Reeve, Brock, 79
Rell, Jody, 145
Rene, Nancy, 240
Renew funding for the program?
  298–302
Retinitis pigmentosa, 35–37, 196–197
Ribas, Antoni, 180–181
Riken, 32
Roberts, Ed, 236

"Roman's Law", 187
Roseman, Achim, 279

Salk vaccine, 112
Samuelson, Joan, 270–271
Schaffer, David, 231–233
Schneer, Jessica, 26–27
Schubert, David, 100–102
Schuele, Birgit, 109
Science Daily.com, 332
Severe combined immune
    deficiency, 2, 7
Sheehy, Jeff, 116–118
Shapiro, Adrienne, 240
Sickle Cell disease, 239–243
Siegel, Bernard, 161–167
Singapore, 156–160
Sipp, Doug, 277–280
Soler, Larry, 313
Solomon, Susan, 207
Spinal Muscular Atrophy (SMA),
    122–124
Srivastava, Deepak, 61–63
Stanford Initiative to Cure Hearing
    Loss, 20–21
Stanko, Steve, 13–14, 16
Star Wars (Movie), 87
Starzl, Thomas, 66–67
States with stem cell research
    programs, 139–144
Stemcellbattles.net, 332
"STEM CELL BATTLES: Proposition
    71 and Beyond", 333
Strong, Gwendolyn, 122–123
Studer, Lorenz, 206–208
Summer Program to Advance
    medical Research Knowledge
    (SPARK), 45

Tarzan, (Movie/Comic), 9
Texas, Cancer Prevention Research
    Institute (CPRIT), 140
Thomas, Jonathan, Chairperson of
    Board of Directors, 46, 117–118,
    199
Thompson, Leslie, 71–72
Tian, Cindy, 148
Tomb of the Unknown Soldier,
    126
Torres, Senator Art, 45
Tourism, stem cell, 277–280
Treves, Avi, 164
Trump, President Donald J., 59–60,
    103–104

UC Berkeley, 229–234
UC Davis, 72–74
Undocumented scientists, 281–283
Urinary incontinence, 259–262

Valley of Death (FDA testing),
    137–138
Vascular endothelial growth factor,
    15
Verne, Jules, 33–34, 37
Vertes, Alain, 164
Viacyte, Inc., 88, 91–92
Villa, Yimi, 193, 301–202
Villafranca, Ernest, 101

Wallace, Milton B., 146
Walsh, Craig, 211–213
Wang, Xiaofang, 142
Wang, Xinnan, 109
Washington, state stem cell program,
    143
Weissman, Irv, 55

Wernig, Marius, 25–26

Whale sharks, 161

Whittaker, Brenden, 257–258

Wing Chun, 285–287

Winokur, Dianne, 246–247

Women in Science, importance of mentoring, 287–292

World Stem Cell Summit, 161–167

Wu, Joseph, 61–63

Yale New Haven Hospital, 147

Yamanaka, Shinya, 26, 297, 299

Yang, Jerry, 147–148

Yeo, Phillip, 156–157

Zika virus, 112, 128–129

Zon, Leonard, 308

www.ingramcontent.com/pod-product-compliance
Lightning Source LLC
Chambersburg PA
CBHW061622220326
41598CB00026BA/3843